COMMUNITY EMPOWERMENT THROUGH RESEARCH, INNOVATION AND
OPEN ACCESS

PROCEEDINGS OF THE 3RD INTERNATIONAL CONFERENCE ON HUMANITIES AND SOCIAL SCIENCES (ICHSS 2020), MALANG, INDONESIA, 28 OCTOBER 2020

Community Empowerment through Research, Innovation and Open Access

Edited by

Joko Sayono & Ahmad Taufiq
Universitas Negeri Malang, Indonesia

Luechai Sringernyuang
Mahidol University, Thailand

Muhamad Alif Haji Sismat
Universiti Islam Sultan Sharif Ali, Brunei Darussalam

Zawawi Isma'il
Universiti Teknologi Malaysia, Malaysia

Francis M. Navarro
Ateneo De Manila University, Philippines

Agus Purnomo & Idris
Universitas Negeri Malang, Indonesia

CRC Press
Taylor & Francis Group
Boca Raton   London   New York   Leiden

CRC Press is an imprint of the
Taylor & Francis Group, an **informa** business
A BALKEMA BOOK

CRC Press/Balkema is an imprint of the Taylor & Francis Group, an informa business

© 2021 selection and editorial matter, the Editors; individual chapters, the contributors

Typeset by MPS Limited, Chennai, India

Although all care is taken to ensure integrity and the quality of this publication and the information herein, no responsibility is assumed by the publishers nor the author for any damage to the property or persons as a result of operation or use of this publication and/or the information contained herein.

Library of Congress Cataloging-in-Publication Data
A catalog record has been requested for this book

Published by: CRC Press/Balkema
                 Schipholweg 107C, 2316 XC Leiden, The Netherlands
                 e-mail: enquiries@taylorandfrancis.com
                 www.routledge.com – www.taylorandfrancis.com

ISBN: 978-1-032-03819-3 (Hbk)
ISBN: 978-1-032-03820-9 (Pbk)
ISBN: 978-1-003-18920-6 (eBook)
DOI: 10.1201/9781003189206

Table of contents

Preface

The Third International Conference on Humanities and Social Sciences (ICHSS) 2020 was held on the campus of Universitas Negeri Malang in East Java, Indonesia on October 28, 2020. This Conference was organized by Social, Humanities, and Tourism Centre (PSP), Institute for Research and Service of Engagement Universitas Negeri Malang. There were five main speakers in this conference including Yuliari Peter Batubara, MBA., Ministry of Social Affairs of the Republic of Indonesia, Assoc. Prof. Haris Abd Wahab from University of Malaya, Assoc. Prof. Chang Chew Hung from Nanyang Technological University, Singapore, Dr. Muhamad Alif Haji Sismat from Universiti Islam Sultan Sharif Ali, Brunei Darussalam, and Prof. Yazid Bustami from Universitas Negeri Malang, Indonesia.

The committee received 135 papers and 65 papers were accepted to be presented. Participants from various universities in Indonesia and some other countries made the conference truly international in scope. Of the total number of presented papers, 30 papers were accepted to be included in the proceedings published by CRC Press / Balkema, Taylor & Francis Group.

Generous support for the conference was provided by the Institute for Research and Community Engagement, Universitas Negeri Malang, and Rector Universitas Negeri Malang for financial support to the event. The 3rd ICHSS 2020 was successfully held and we invite the presenters around the world to participate in the next ICHSS that will be held in 2022. Finally, given the rapidity with which social science and humanities developed, we hope that the future ICHSS will be as stimulating as indicated in this proceedings volume.

Acknowledgement

All papers in this book are the final version of manuscripts that were presented at the 3rd International Conference on Humanities and Social Science (ICHSS), held on October 28, 2020 at Malang, Indonesia. We would like to thank the Rector, Institute for Research and Community Engagement (IRCE) Universitas Negeri Malang, and all of the staff for their great support. On behalf of the organizers, authors, and readers, we also wish to express thanks to all keynote speakers and reviewers for their hard work, time, and dedication to make this conference a success. Their effort could maintain the high standards of all papers in this book. Our thanks also go to the participants and presenters of the conference. Finally, many thanks are given to all committee and for all persons who helped and supported this conference.

The organizers wish to apologize to the presenters who could not publish his/her paper in this conference proceeding. Our apology also goes to all participants for all shortcomings in this conference. See you in the next 4th ICHSS at Universitas Negeri Malang.

Malang, January 22, 2021

Organizer of ICHSS
Universitas Negeri Malang

Scientific committee

Syamsul Bachri, Ph.D
Universitas Negeri Malang, Indonesia

Dr. Ahmad Taufiq
Universitas Negeri Malang, Indonesia

Prof. Ir. Achmad Subagio, M.Agr, Ph.D
Universitas Negeri Jember, Indonesia

Prof. Dr. Darni, M.Hum
Universitas Negeri Surabaya, Indonesia

Assoc. Prof. Dr. Zawawi Isma'il
Universitas Malaya, Malaysia

Dr. Alicia Schrikker
Universiteit Leiden, Belanda

Assoc. Prof. Dr. Luechai Sringernyuang
Mahidol University, Thailand

Prof. Dr. Warsono, M.S
HISPISI, Indonesia

Organizing committee

Advisory Board
Prof. Dr. Ah. Rofi'uddin
Universitas Negeri Malang, Indonesia

Chairman:
Dr. Joko Sayono, M.Pd, M.Hum
Universitas Negeri Malang, Indonesia

Secretary
Agus Purnomo,S.Pd, M.Pd
Imamul Huda Al Siddiq, M.Sosio
Universitas Negeri Malang, Indonesia

Treasurer
Prihatini Retnaningsih, S.E.
Universitas Negeri Malang, Indonesia

Technical Chair
Ronal Ridhoi, S.Hum, M.A.
Lisa Sidyawati S.Pd., M.Pd
Adi Prasetyawan, S.Sos, M.A.
Indarti Adininggar, A.Md
Dimas Arif Dewantoro, M.Pd
Universitas Negeri Malang, Indonesia

Publication
Dr. Ahmad Taufiq, S. Pd., M. Si
Idris, S.S., M.M.

Community Empowerment through Research, Innovation and Open Access – Sayono et al (Eds)

Empowering translation students through the use of digital technologies

M.A.H. Sismat*

Sultan Sharif Ali Islamic University, Bandar Seri Begawan, Brunei Darussalam

ABSTRACT: Digital technologies are widely used in various settings, including in translator training. The advancement of digital technologies, particularly translation technologies, has altered both translator training and professional setting. The present study investigates how these technologies can empower students in their learning environment. A group of 21 undergraduate translation students at Sultan Sharif Ali Islamic University conducted a series of 12-week media translation projects, consisting of social media translation and subtitling. The study results were made based on their project reports and a questionnaire to determine their attitudes towards technologies, such as Aegisub and Memsource. Based on the findings, the students showed positive attitudes towards projects and technologies. Most of them see improvements in their confidence. The projects prepared them with sufficient translation and technical skills to become professional translators, including critical thinking and teamwork skills.

Keywords: translation, digital technologies, learning environment

1   INTRODUCTION

Empowerment is a powerful word, which may imply different meanings depending on the context. Cambridge Dictionary (n.d.) defined the term as "to give someone the official authority or the freedom to do something." If students are put into context, it generally means to authorize the students to do something. However, this definition remains vague and may be misinterpreted. It lacks the intended meaning that the present study tries to imply. It agrees more with Collins Dictionary (n.d.), defining it that "to empower someone means to give them the means to achieve something." This definition is more suitable for this study's context, as it intends to investigate how digital technologies can empower translation students. Lightfoot (1986) mentioned four elements in decision making when defining empowerment: autonomy, choice, responsibility, and participation. These four elements are crucial suitable when empowering students in a learning environment. Even though they are free and encouraged to make decisions for themselves, they must be responsible for every action they make.

Recent years have seen significant changes in the translation industry as the demands for translation and localization services have been rapidly growing. The increasing demands have led to the importance of producing good quality translations in a short time. As a result, there has been a shift in the translation profession. It now requires technical skills to meet the ever-increasing demands of clients. Machine translation (MT) was the first translation technology introduced in 1954 and designed for the Georgetown-IBM Experiment to translate from Russian into English (Hutchins 2004). However, the effort stopped for a long time as they hit a dead-end. Later in the second half of the 1970s and early 1980s, studies on MT were revived. Several MT systems, such as Systran and Logos, emerged for general application even though the dictionaries they used were modified for specific subject areas (Hutchins 2007). Since then, MT systems have evolved from

---

*Corresponding author: alif.sismat@unissa.edu.bn

DOI 10.1201/9781003189206-1   1

using rule-based to the current neural approach, translating more than 100 languages for various purposes.

Another well-known translation technology is the computer-assisted translation (CAT) tool. In 1990, the first CAT tool was first introduced by TRADOS, which is later acquired by a language service provider, SDL, and now known as SDL Trados. Compared to MT systems, CAT tools typically consist of main features, such as Translation Memories (TMs), Terminology Management Database (TermBase), alignment. However, as times change, CAT tool developers have incorporated numerous features, such as MT systems and cloud-based management systems. These upgrades have increased translators' productivity and quality (Koponen, Aziz, Ramos & Specia 2012; Koponen & Salmi 2015; Sismat 2019) and improved attitudes towards the tools (Guerberof 2012; Alotaibi 2014; Çetiner 2018).

Over the past decade, CAT tools and MT systems have become "an integral part of translator training and professional practice" (Sismat 2016). The new normal requires translators to adapt post-editing (PE) practice in their work. Such changes have also altered translation training. Translation programs offer more than just learning and applying translation theories, methods, and strategies; but they even offered technology-related modules to train students to become tech-savvy professionals.

The current COVID-19 pandemic has also impacted the translation and language industry as it has forced more translators to work remotely from home. However, this is not a new normal for freelance translators who have been working at home before the first outbreak. Translators who have technological skills remain on the top. The pandemic has also changed the contents to be translated. For example, media contents on various platforms, such as Netflix, YouTube, and social media, have quickly attracted more users as people stay at home due to movement restriction order and lockdown since March 2020. The surge in the number of online users has led to translating a more considerable amount of media content.

Socially, more established terms have emerged and are commonly used in the new normal, such as social distancing, contact tracing, lockdown, shutdown, and self-quarantine. These terms are historically not fresh as they were popular in past events. The term 'coronavirus' itself was first used in 1968 during the H3N2 virus pandemic. In the 17th century, the term 'self-quarantine' was used during an outbreak of bubonic plague. Coronavirus neologisms have also been widely used and may sound alien to many people. For example, colonials describe the generation born during the COVID-19 pandemic. Covidiot is a combination of the terms 'covid' and 'idiot,' representing people who break the government guidelines in combating the COVID-19 crisis, while infodemic is a combination of the terms 'information' and 'pandemic' to tell the spread of information that causes speculations, panic or anxiety. Most of the terms, as mentioned above, can be tricky to translate and may sound alien to many people who speak other languages. For example, the word 'social distancing' is translated to Malay as *'penjarakan sosial'*. This term was 'new' and odd to Malay people, particularly in Brunei, but now it is widely accepted and used. However, not all of these terms are loosely translated in Malay. Instead, these terms were also borrowed in Malay to make it sound more familiar, such as *infodemik, asimtomatik,* and *generasi coronial.*

Due to the shifts in the contents and demands for technical skills in the translation industry the translator training has also been altered as computer-assisted translation (CAT), machine translation (MT), and media translation have become core modules in most higher education institutions that offer translation programs, including Sultan Sharif Ali Islamic University in Brunei Darussalam, or locally known as Universiti Islam Sultan Sharif Ali or UNISSA.

In these advanced modules, translation students were given the opportunity to conduct their translation projects to empower them to use their critical thinking skills in making decisions. Therefore, the present study aims to investigate the roles of digital technologies in empowering translation students at UNISSA. Therefore, this study will answer the following research questions:

RQ1: What are the translation students' attitudes towards the use of digital technologies?
RQ2: How can digital technologies empower the students learning process?

2 METHODS

2.1 *Participants*

The present study involved 21 undergraduate students who were majoring in Arabic language and translation at UNISSA. The students were Malay native speakers and typically translated between three languages: Malay, Arabic, and English. They had previously studied the basic knowledge of translation methods and strategies, and most of them had learned English and Arabic for more than 12 years.

2.2 *Translation projects*

For this study, the data were collected from translation projects in one of the core modules: 'Media Translation.' The module includes audiovisual translations, which required the students to conduct subtitling projects, and social media translation, which involves the use of CAT tools and MT engines.

A series of six translation projects were conducted within 12 weeks. In each project, the students play three different roles: project manager, translator, and proofreader. Before running these projects, the students were given hand-outs and a series of hands of practice to familiarize them with the subtitling software programs. They must use the previously mentioned translation technologies to translate social media content for the first three projects and then focus on the remaining three projects' subtitling projects.

Each project was usually carried out within two weeks. As project managers, the students must find texts and contents suitable for their team members to translate. The translators then received the materials and began with the translation process. However, as part of the subtitling projects, the translators must complete the transcription process first before they could translate them. Once translated, the translators had to submit both transcript and translation to the project managers, who would pass the translations to the proofreaders to analyze and correct any translators' errors. Once edited, the groups could begin with the subtitling phase, which requires precision, critical thinking, and teamwork skills. At the end of each project, the project managers had to present their final products and share their learning experience, particularly finding solutions to any difficulties they faced in their projects.

Also, translators and proofreaders had to submit a report on their work to share their working ethics and solutions to solve any difficulties. These reports reflect their level of critical thinking skills in decision making. Apart from the report, they had to also submit invoices for their service as practiced in the translation industry. This exercise was incorporated into the projects to give an insight into the professional setting and increase their interest in becoming professional translators.

2.3 *Digital technologies*

The present study uses various digital technologies as contents and assisting tools. In social media translation tasks, project managers were allowed to choose any social media content. Most of them chose popular platforms such as Instagram, Facebook, and Twitter, mainly advertisements, news, and entertainment. For the social media translations, they must translate the contents using a cloud-based CAT tool, called Memsource, which incorporates MT engines in the program. On the other hand, the subtitling tasks required them to use a free, open-source subtitle editor tool, called Aegisub. A wide range of videos was translated for these tasks, such as songs, cartoons, film trailers, motivational videos, and educational videos, obtained from different sources, such as YouTube, Facebook, and other media websites. Also, translators and proofreaders must submit a report on their work to share their working ethics and solutions to solve any difficulties. These reports reflect their level of critical thinking skills in decision making. Apart from the report, they must also submit invoices for their service as practiced in the translation industry. This exercise was incorporated into the projects to give an insight into the professional setting and increase their interest in becoming professional translators.

2.4 *Data analysis*

A questionnaire was designed to determine the students' attitudes towards digital content and technologies to answer the research questions. The questionnaire was divided into three sections. The first section consisted of subjective questions regarding the project information as follows:

a. What type(s) of content you used in the social media translation projects?
b. Which one of them do you prefer best? Why?
c. What type(s) of content you used in the subtitling projects?
d. Which one of them do you prefer best? Why?
e. What is the most difficult process in subtitling? Why?

The second and third sections consisted of 5-point Likert scale questions to validate the following statements:
Section 2: Social media translation

a. I enjoy translating social media content.
b. Social media contents are easy to translate.
c. Memsource is easy to use and suitable for translating social media content.
d. Raw Machine Translations generated from Memsource Translate are editable and usable.
e. The media content influenced the quality of the translation.
f. The projects improved my confidence as a translator.

The projects encouraged my creativity.
Section 3: Subtitling

a. I enjoy subtitling.
b. Subtitling is a delicate work, which requires both translation and technical skills.
c. Memsource is suitable for translating transcripts.
d. Aegisub is easy to use and suitable for subtitling.
e. The media content influenced the quality of the subtitles.
f. The projects improved my confidence in subtitling.
g. The projects encouraged my creativity.

As previously mentioned, the projects required the students to submit a report depending on their role. Their translations and reports were examined to see whether they acquired crucial skills, such as critical thinking and teamwork skills, in solving translation problems and technical skills.

3 RESULTS AND DISCUSSION

3.1 *Project information*

Table 1 represents Questions 1 and 2 in Section 1 of the questionnaire regarding the contents translated in the social media translation projects and the preferred contents. Over one-third (8 out of 21) of the students translated news posts, while others translated infographics (5 students), advertisements (4 students), and healthcare posts (4 students). However, when asked about their preferred content, almost half of them chose infographics. They stated that the infographics consist of fascinating subject matters, such as local economic developments, languages, and Covid-19 statistics. Meanwhile, one-third of them chose news because they are current and easy to understand, such as the covid-19 pandemic, online learning, elections, and local news reports.

The students were then asked similar questions on the subtitling projects. Table 2 indicates that the students subtitled various videos such as comedy, songs, cartoons, trailers, motivational and educational videos. When asked their preferred contents in Question 4, 7 of 21 students chose comedy videos, stating that comedy videos are fun to translate. Therefore, it makes the subtitling process less stressful. 5 out of the students chose cartoons, stating that subtitling animated films are easy to translate because they had previously watched the movie repeatedly when they were

Table 1.   Types of contents and preferences in social media projects.

Type of content	Q1	Q2
News	8	7
Advertisement	4	2
Infographics	5	10
Healthcare	4	2

Table 2.   Types of contents and preferences in the subtitling projects.

Type of content	Q3	Q4
Songs	2	2
Animation	5	7
Cartoons	3	5
Trailers	3	3
Motivational videos	4	2
Educational videos	4	2

Table 3.   Results of section 2 of the questionnaire.

Statement	Strongly agree	Agree	Neutral	Disagree	Strongly disagree
1	7	10	4	–	–
2	18	3	–	–	–
3	–	15	6	–	–
4	–	16	5	–	–
5	4	13	3	1	–
6	2	14	5	–	–
7	3	16	2	–	–

kids. Others who preferred songs and trailers said they are of their interests, such as singing and playing video games.

In response to Question 5, most of them stated that both spotting and line break is the most difficult processes as the former requires precision. The latter is limited to 42 characters in two lines per subtitle. They had to rephrase some of the words to ensure that the subtitle does not exceed the character limit. Exceeding them would affect their readability. Some of them also stated that the transcription process is the most challenging as it requires excellent listening skills.

3.2   *Social media translation*

Table 3 shows the second part of the questionnaire's results regarding the students' experience and attitude towards social media translation projects. Regarding Statement 1, the results indicate that most of them enjoyed the project. A possible explanation for this may be related to Statement 2, indicating that 18 out of 21 students found the social media contents are easy to translate while the other three students remained neutral. When examining their reports, some students stated that hashtags are tricky to translate as the translation must also be as catchy as the original. For example, the hashtag *#banar2brunei* was challenging to translate to Arabic as it could mean بروناي الحقيقي (Truly Brunei) or بروناوي حقيقي (a True Bruneian). However, the translator chose the latter as it is more suitable for the given context.

In response to Statement 3, most of the students (71.4%) agreed that Memsource is easy to use and suitable for translating social media contents. A possible explanation for this is that CAT tools typically preserve the formatting of the file. Therefore, it reduces the number of efforts in designing and formatting the content. Furthermore, when examining their response to Statement 4,

Table 4. Results of section 3 of the questionnaire.

Statement	Strongly agree	Agree	Neutral	Disagree	Strongly disagree
1	10	11	–	–	–
2	2	14	5	–	–
3	-	8	9	4	–
4	-	14	7	–	–
5	4	15	2	–	–
6	3	15	3	–	–
7	10	9	2	–	–

16 students found the raw MTs generated from Memsource Translate were editable and usable when translating the social media contents. This statement may be a possible explanation for the previous statement regarding the easiness and suitability of Memsource for translating the contents.

In response to Statement 5, most of the students agreed that the contents influenced their translations' quality as they stated in their report that it is easy for them to translate the exciting contents. They must be knowledgeable of the subject matter. This statement is linked to Statement 6 regarding their confidence after carrying out the projects. 16 out of 21 students agreed that the projects improved their confidence as a translator while the other five remained neutral, indicating that content and subject matter contribute to improving their confidence level. Similarly, in Statement 7, most students agreed that the projects encouraged their creativity, suggesting that social media content and translation projects can empower them. These exercises boost their confidence, creativity, and critical thinking skills, as reflected in their reports.

3.3 *Subtitling*

For the subtitling project, students were asked similar questions to the previous questions. 4 shows the last section of the questionnaire's results regarding the students' experience and attitude towards subtitling projects.

Similar to the previous section, students were first asked if they enjoy subtitling. Based on their response to Statement 1, shown in 4, all of them enjoyed subtitling. However, most of them also understood that subtitling processes are delicate and require both translation and technical skills (Statement 2). Their reports also stated that they helped each other when they faced transcription and subtitling difficulties, indicating their excellent teamwork skills (Çetiner 2018).

In response to Statement 3, only 8 (38.1%) out of 21 students agreed that Memsource is suitable for translating transcripts. 9 students remained neutral, while 4 of them disagreed. A possible explanation for this is that based on their reports, and some stated that they preferred translating the transcripts without Memsource because its MT, Memsource Translate, could not provide editable and usable translations for some contents. Particularly, songs and films may contain idiomatic expressions and metaphors that MT cannot translate directly. For example, when translating the idiomatic expression 'it's raining cats and dogs' into Malay, Memsource Translate rendered it as '*hujan kucing dan anjing*', which completely does not make any sense. Another example is the term chicken as a metaphor for a coward. However, Memsource Translate rendered 'You're a chicken' literally as '*Awak ayam*', which sounds comical as it is not the intended meaning.

When asked about the easiness and suitability of the editing tool (Statement 4), two-thirds of the students agreed that Aegisub is easy to use and suitable for subtitling projects. Some of them stated that the 'shift times' feature enhances the subtitling process. It allows users to modify the start/end timestamps by time or frames. They also found that the software interface is user-friendly and straightforward. In response to Statement 5, most students agreed that the contents also influenced the subtitles' quality. They stated in their reports that different types of content require different approaches to translating them. Even though they stated that they enjoyed subtitling, expressions in some content, such as songs and films, require deeper understanding and knowledge of the equivalent terms and phrases in the target language and culture (Alotaibi 2014; Hutchins 2007).

Despite the complexity of the subtitling linguistically and technically, most students (85.7%) agreed that the subtitling projects boosted their confidence as translators (Statement 6). A possible explanation for this is that some of them reported that Aegisub was challenging to use at first. However, they slowly familiarized themselves with the tools and found it easy to use after a while. The technical skills they acquired increased their confidence level and prepared them for professional settings. In response to Statement 7, most students (90.5%) also believed that the projects encouraged their creativity, reflected in their presentation skills. In addition to the main contents, project managers were competitive in producing introduction videos as their creative way to introduce their projects and team members. The videos also surprisingly looked professional and were of high quality.

4 CONCLUSION

The overall findings suggest that digital technologies can empower translation students through translation projects to some extent. Despite facing some difficulties, students managed to use their critical thinking skills when carrying out the projects technically and linguistically. Based on the results, Memsource seems to be more suitable for translating social media contents than translating the videos' transcript, due to the incapability of Memsource Translate in providing equivalent terms and phrases, particularly idiomatic expressions and metaphors in songs and films. The complexity of the contents also influenced the quality of the final products. Apart from the equivalence issues, the contents also need to be interesting to make the translation and subtitling processes more fun and less stressful. As a result, most students also displayed positive attitudes towards digital content and technologies in social media translation and subtitling projects. They also saw improvement in their confidence because they prepared them with the necessary skills to become professional translators and subtitlers. Surprisingly, the students also showed their competitiveness and teamwork skills, mainly when presenting their projects.

REFERENCES

Alotaibi, H.M., 2014. Teaching CAT Tools to Translation Students: An Examination of Their Expectations and Attitudes. *Arab World English Journal*.
Cambridge Dictionary, n.d. *Empower Meaning in The Cambridge English Dictionary*.
Çetiner, C., 2018. Analyzing the attitudes of translation students towards cat (computer-aided translation) tools. *Journal of Language and Linguistic Studies*, *14*(1), pp.153–161.
Collins Dictionary, n.d. *Empower Meaning In The Cambridge English Dictionary*. [online] https://www.collinsdictionary.com.
Guerberof, A.A. 2012. *Productivity and quality in the post-editing of outputs from translation memories and machine translation* (Doctoral dissertation). Universitat Rovira I Virgili, Tarragona, Spain.
Haji Sismat, M. A. 2016. *Quality and productivity: A comparative analysis of human translation and post-editing with Malay learners of Arabic and English* (Doctoral dissertation). University of Leeds, Leeds, United Kingdom.
Haji Sismat, M. A. 2019. Inverse Translation Quality: A comparative analysis between human translation and post-editing. *Journal of Arabic Linguistics and Literature*, 2, 91–105.
Hutchins W.J. 2004. The Georgetown-IBM Experiment Demonstrated in January 1954. In: *Frederking R.E., Taylor K.B. (eds) Machine Translation: From Real Users to Research. AMTA 2004. Lecture Notes in Computer Science*, 3265. Springer, Berlin, Heidelberg. https://doi.org/10.1007/978-3-540-30194-3_12
Hutchins, J. 2007. Machine translation: A concise history. *Computer aided translation: Theory and practice*, *13*(29–70), p.11.
Koponen, M., Aziz, W., Ramos, L., & Specia, L. 2012. Post-editing time as a measure of cognitive effort. In *Proceedings of WPTP*, 11–20.
Koponen, M., & Salmi, L. 2015. On the correctness of machine translation: A machine translation post-editing task. *The Journal of Specialised Translation*, 23, 118–136.
Lightfoot, S.L. 1986. On Goodness in school: Themes of empowerment, *Peabody Journal of Education*, 63:3, 9–28.

Community Empowerment through Research, Innovation and Open Access – Sayono et al (Eds)
© 2021 Copyright the Author(s), ISBN 978-1-032-03819-3

The role of university on economic development in heritage tourism area in Malang City, Indonesia

V.A. Qurrata*, S. Merlinda, V. Purnamasari
Universitas Negeri Malang, Malang, Indonesia

ABSTRACT: Community service activities in revitalizing religious tourism were carried out to increase tourist numbers, mostly visits by young generations, expanding the community's economy and introducing history to visitors. Malang City's relatively high tourism potential must be the point of attention of all social elements. University is expected to be a role model in analyzing solutions to problems to create a prosperous and independent society through tourism. Activities were carried out using the focus group discussion method to socialize and prepare for development. We then assisted in developing and implementing training and evaluation of performance. The service contributed to iconic photo spots and parks around *Mbah* Honggo's grave, direct marketing strengthening movement through *Mbah* Honggo's historical posters, and social media training (Google My Business). The evaluation results provide concluded that the number of tourist visits is increased so that the surrounding community's economy is also developed. This also served as additional historical knowledge for visitors.

Keywords: economics development; *Kampoeng* heritage; tourism area; religious tourism

1 INTRODUCTION

Malang City is one of the tourist areas in East Java, Indonesia, visited for tourism. According to data from the Malang City Tourism Outlook (2019), the tourism sector contributes 25.56 percent of PAD to Malang City. The number of tourists in 2019 amounted to 5.170.523 and foreign tourists of 16.286 (BPS 2019). It is an opportunity and a challenge to make maximum use of the tourism potential in Malang, especially in the heritage tourism area in the era of digitalization. The development of the internet is one of the new tools in capturing and expanding potential customers, specifically aimed at millennials tourism. Thus, tourism development is directed to be a mainstay sector capable of helping or competing activities in the economic sector and other related industries (Sutiarso 2017). Community service activities in the form of revitalizing religious tourism are carried out to increase tourists' numbers, mostly visits by young generations, expanding the community's economy and introducing history to visitors.

This activity is intended for managers and the religious tourism community of *Mbah* Honggo, who are precisely located at *Kampoeng* Heritage Kajoetangan as many as ten people. In its development, the religious tourism community of *Mbah* Honggo faces problems due to the inadequate facilities available. It is a crucial problem because, in the era of millennials tourism, attractive facilities are used as capital to absorb potential consumers as a positive impact of the internet in the tourism sector, namely low promotional costs (Rusdi 2019) and the uneven information regarding the existence of religious tourism in *Kampoeng* Heritage.

Therefore, to solve partner problems, the university provides an analysis of issues and solutions to partners in *Kampoeng* Heritage Kajoetangan. The activities were carried out using Focus Group

*Corresponding author: vika.annisa.fe@um.ac.id

DOI 10.1201/9781003189206-2

Discussion to socialize and prepare for development, and we also assisted in the implementation of development and implementation training and evaluation of performance. The service contributed to iconic photo spots and parks around *Mbah* Honggo's grave, direct marketing strengthening movement through *Mbah* Honggo's historical posters, and social media training through Google My Business. Based on the evaluation and monitoring of this activity, it concluded that the number of tourist visits is increased so that the surrounding community's economy is also developed and adds as historical knowledge for visitors.

2 METHODS

In developing *Mbah* Honggo, religious tourism was carried out in three strategic steps. The first is the Focus Group Discussion method to socialize and prepare for development. The FGD model was carried out to identify the problems and the size of the tourism potential (Pratama 2019). In this case, the FDG included a tourism community, representatives of local and regional leaders, and community service programs. Second, we assisted in the implementation of development and implement training. The program was installing a letter sign and improving the marketing strategy for the *Mbah* Honggo Tomb Area. It was carried out with the following stages and methods: preparing an activity plan and collecting data using primary data and secondary data through field observations. Moreover, exploring the site, documentation, and interviews were also done. Literature study on the concept of suitable letter sign design and marketing strategies is required to attract tourists, especially the younger generation, to visit there, qualitative descriptive data analysis, and installation of letter signs and online marketing training for tourism area managers. Finally, evaluation of the implementation was done.

3 RESULTS AND DISCUSSION

As a part of cultural tourism, heritage tourism has a significant attraction that proliferates (Franch et al. 2017). The existence of heritage tourism can increase opportunities and interactions between residents and visitors to increase income (Nicholas & Thapa 2010). Therefore, the citizen's role in the development of heritage tourism is crucial in its progress and development. One of the population's roles in developing tourist areas is services, such as communication skills in foreign languages, and tourism maintenance services such as maintaining advice and infrastructure (Zhang & Stewart 2017). Resources in tourism include an environmental component, either physical or social, which provides the infrastructure to attract tourist visits (Bucurescu 2012; Cooper & Hall 2008; Darabseh et al. 2017). The results and discussion of community service programs in the form of focus group discussions and assistance in the form of facilities at *Mbah* Honggo Religious Tourism and training for the community were divided into three stages. The SWOT analysis of the *Mbah* Honggo religious tourism area is shown in Figure 1.

SWOT analysis presented in the Figure above shows the real condition of *Kampoeng* Heritage Kajoetangan. Based on the explanation, several weaknesses exist in the place's internal management, and most of them are caused by the lack of facilities. While facing the internal weakness, the management also faces other external challenges such as *Kampoeng* Heritage Kajoetangan is threatened by other tourism places in Malang. The most significant issue in this problem includes limited promotion held by the management.

On the other hand, there is a massive strength that the *Kampoeng* Heritage Kajoetangan has, such as strategic location and sentimental object, which can attract tourists. Opportunities will empower and support the local communities' strength if they put a significant effort to develop this area. From the SWOT analysis data, a strategy that can be taken from the partners' problems can be represented in Figure 2.

The SWOT analysis be conducted based on the community's internal activities while adjusting with the external fact that happened in real life. Using the SWOT analysis internally, we can

A. Strength	B. Weaknesses
• S1: has the power as a place of religious and historical tourism • S2: has the strength as a comfortable place to travel • S3: located in the middle of the city, so it is easy to access	• W1: There are still a few supporting facilities • W2: no letter sign can also be used as a visitor photo spot • W3: there is a management conflict
C. Opportunities	D. Challenges
• O1: tourists, especially the younger generation, like Instagram-able tourist attractions • O2: tourists looking for tours that are cheap and easy to reach	• Q1: lack of promotion on social media • Q2: Many thematic tourist locations have sprung up in Malang

Figure 1. SWOT analysis of the *Mbah* Honggo religious tourism area.

SWOT Strategy:

SO1: increase the number of iconic photo spots for visitors
SO2: there are superior regional products that can be sold
ST1: do promotions through social media
ST2: visitors are asked to upload photos there along with a hashtag
WO1: add a letter sign that can be used as well as lighting in the place
WO2: management involving all RW residents covering the area
WT1: specialized in thematic village areas
WT2: made additional café facilities and regional souvenirs

Figure 2. SWOT strategy of the *Mbah* Honggo religious tourism area.

maximize the strength and minimize the weaknesses. Externally, we could exploit the challenges and opportunities available to support a better vision in the future. Using the SWOT Strategy above, we focus on several activities that combine activities using strength to challenge the opportunities, use strength to minimize the threat, minimize the weakness by using the opportunities, and minimize the weaknesses based on the threat in reality.

This analysis decides to improve facilities and strengthen marketing strategies through social media. This decision is also assessed based on the essential needs, according to managers and the community. With an attractive photo spot, the hope is that tourist visits will increase (Kuenzi 2008). The coordination process with the local area head, community group, and the site management resulted in planning activities, including budget planning and costs adjusted to the RAB. The coordination results concluded that this tourism revitalization's priority needs were the construction of iconic photo spots accompanied by lighting (neon boxes), strengthening direct marketing through foreign languages in serving tourists, and strengthening promotions through Google My Business.

In this program, residents would assist in the construction of iconic photos and strengthen direct marketing carried out by the team. One of the things that were done by the group was to find, collect, and summarize *Mbah* Honggo's historical data. The summary is also discussed and validated by the local community watch, which will then be displayed in the historical tourism area. Furthermore, the community service team's revitalization process is monitored by the regional area head. After handing over the aid, the community groups, then the community, made improvements to tourism facilities, namely making iconic photos that also serve as lighting for the area. Then, our team also assists in improving tourism support facilities, as stated in the following figure.

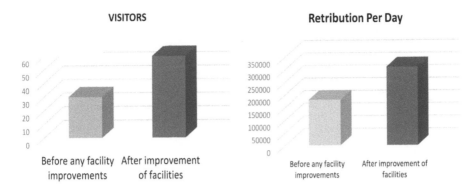

Figure 3.   Visitors (left side) and retribution per day (right side).

Furthermore, the development of iconic photo spot development was in the development process. After the revitalization was carried out, especially on iconic photos, another thing that was also done was direct marketing education. Education direct marketing is implemented in English strengthening for all residents, including community watch and women's groups, to facilitate marketing and improve the performance of more excellent tourism services. Moreover, a form of direct marketing education is the inclusion of *Mbah* Honggo's history in the form of posters, which are packaged in two languages, namely Indonesian and English. It is due to the convenience of guides in explaining and foreign tourists in understanding the history and origin of *Mbah* Honggo.

After the design of *Mbah* Honggo's History pamphlet was compiled, it was printed and handed over to the community watch group and the local head area, witnessed by the local community and representatives of the Malang City Tourism Office. The final implementation of this activity is social media promotion training. The presence of social media nowadays has become a significant need (Fandeli 2001). All age groups have used social media for communication needs and business needs (Wolah 2016). The media used was the creation of Google My Business. Google My Business is a platform from Google, where business owners can provide detailed business information. When someone searches on a search engine, the search results will appear complete with a map of directions, making it easier for potential customers to find Google locations. Besides, through this feature, it will be easier for business owners to create and update listings so that their business will stand out more. For entrepreneurs and businesspeople, Google My Business is very useful for promoting a business, and from the satisfactory promotional results, business managers can increase the company's sales. The detail of visitor numbers and retribution numbers per day can be seen in Figure 3.

The development of *Mbah* Honggo religion tourism shows positive growth, an increase in tourist visits from before and after the facility's construction. It also provides a multiplier effect of an increase in the number of fees. These results align with research in developing a tourist village in Tulungrejo Village by Amalia et al. (2018). This research shows new economic activities that open jobs and additional income for the surrounding community. This effect also affects farmers in increasing the selling value of apples and research conducted by Wijayanti (2017), in descriptive survey research that analyzes the impact of developing the tourism village of Kembangarum on the community's economy.

4   CONCLUSION

Opportunities and challenges to exploit the tourism potential in Malang City, especially in the heritage tourism area in a digital age, are very crucial and require the right strategy. In resolving partner problems, namely supporting facilities for heritage tourism villages, we are revitalizing historical tourism by creating iconic photo spots. The landmarks were built as iconic photo spots that function as grave markers and visitors' sites to capture moments in the Religious Tourism area of *Mbah* Honggo's Tomb. In addition to physical revitalization, we also conduct search engine marketing training to facilitate promotion strengthening and market capture. Then, the strengthening of direct marketing was carried out to support all languages of the citizens, including community watch and women's groups, to provide full services to domestic tourists and foreign tourists. This program can increase the role of universities in increasing the competitiveness of tourist areas. The revitalization of heritage tourism has risen to visitors of up to 100 percent, accompanied by an increase in fees. Thus, the income of the surrounding community can rise through the opening of new jobs.

REFERENCES

Amalia VGA, N., Kusumawati, A., & Hakim, L. 2018. Partisipasi Masyarakat dalam Pengembangan Desa Wisata serta Dampaknya terhadap Perekonomian Warga di Desa Tulungrejo Kota Batu. *Jurnal Administrasi Bisnis*, 61(3), 48–56.

BPS. 2019. Statistik Kunjungan Wisatawan Mancanegara. Accessed on November 24 at 13.39 WIB. Available at: https://malangkota.bps.go.id/subject/16/pariwisata.html#subjekViewTab1

Bucurescu, L. 2012. Assessment of tourism potential in historic towns: Romanian case studies. In International conference: *The role of tourism the territorial development* (conf. Proceeding). Press Universitatea Clugeana, Gheiorgheni.

Cooper, C., Hall, M. 2008. *Contemporary tourism: an international approach*. Butterworth-Heinman London.

Darabseh, F.M., Ababneh, A. & Almuhaisen, F. 2017. Assessing Umm el-Jimal's Potential for Heritage Tourism. *Arch* 13, 460–488. https://doi.org/10.1007/s11759-017-9327-5

Faizal, E., Suprawoto, T., Kurniyati, N. N., & Setyowati, S. (2020). Pengembangan Wisata Tematik Sebagai Rintisan Kawasan Edukatif Ramah Anak. *Jurnal Berdaya Mandiri*, 2(1), 202–214.

Fandeli, C. 2001. *Dasar-Dasar Manajemen Kepariwisataan Alam*. Yogyakarta.

Franch, M., Irimiás, A. & Buffa, F. 2017. Place identity and war heritage: managerial challenges in tourism development in Trentino and Alto Adige/Südtirol. Place Brand Public Dipl 13, 119–135. https://doi.org/10.1057/s41254-016-0019-5

Komariah, N., Saepudin, E., & Yusup, P. M. 2018. Pengembangan Desa Wisata Berbasis Kearifan Lokal. *Jurnal Pariwisata Pesona*, 3(2). https://doi.org/10.26905/jpp.v3i2.2340

M. J. Kuenzi C., 2008. *Nature-Based Tourism*. In: Renn O., Walker K.D. (eds) Global Risk Governance. International Risk Governance Council Bookseries. Dordrecht: Springer.

Nicholas, L., Thapa, B. 2010. Visitor perspectives on sustainable tourism development in the Pitons Management Area World Heritage Site, St. Lucia. *Environ Dev Sustain* 12, 839–857. https://doi.org/10.1007/s10668-009-9227-y

Pitana, I Gde, Gayatri, dan Putu G. 2005. *Sosiologi Pariwisata*. Yogyakarta: Andi Offset.

Rusdi, J. F. 2019. Peran Teknologi pada Pariwisata Indonesia. *Jurnal Accounting Information System (AIMS)*. https://doi.org/10.32627/aims.v2i2.78

Setiadi, Amos., L.A. Rudwiarti. 2020. Penataan Kawasan Wisata Curug Banyu Nibo Desa Sendangsari Kecamatan Pajangan Kabupaten Bantul Daerah Istimewa Yogyakarta. *Jurnal PATRIA* (media online) Vol. 2(1).

Sutiarso, M. A. 2017. *Pengembangan Pariwisata Yang Berkelanjutan Melalui Ekowisata*. Manajemen Kepariwisataan Di Sekolah Tinggi Pariwisata Bali Internasional (STPBI).

T. D. Oonowska M. 2016. Toward a sustainable tourism. In Tourism Management, Marketing, and Development. In: Mariani M.M., Czakon W., Buhalis D., Vitouladiti

Wijayanti, A. 2017. Analisis Dampak Pengembangan Desa Wisata Kembang Arum Terhadap Perekonomian Masyarakat Lokal. *Sarjana Wiyata Tamansiswa* Yogyakarta.

Wiyatiningsih. 2015. Global-Lokal: Kreativitas Meruang Sebagai Strategi Keberlanjutan Desa Wisata Puton Bantul. *Konferensi Nasional II Forum Wahana Teknologi Yogyakarta*, Yogjakarta, 10 Agustus 2015.

Wolah, Ferni Fera Ch. 2016. Peranan Promosi dalam Meningkatkan Kunjungan Wisatawan di Kabupaten Poso. *Acta Diurna*. 5(2), 23–40.

Yuniwati, E. D., Darmawan, A. A., & Firdaus, R. 2020. Eksplorasi Potensi Alami Waduk Menuju Rancangan Wisata Desa Purwosekar Tajinan Kabupaten Malang. *Dinamisia: Jurnal Pengabdian Kepada Masyarakat*, 4(3), 398–407.

Zhang, L., Stewart, W. 2017. Sustainable Tourism Development of Landscape Heritage in a Rural Community: A Case Study of Azheke Village at China Hani Rice Terraces. *Built Heritage* 1, 37–51. https://doi.org/10.1186/BF03545656.

Community Empowerment through Research, Innovation and Open Access – Sayono et al (Eds)
© 2021 Copyright the Author(s), ISBN 978-1-032-03819-3

Formulation of curriculum moderation on religious mentoring: Case study from three public universities in Indonesia

A.M. Nasih*, A. Sultoni, T. Thoriquttyas
Universitas Negeri Malang, Indonesia

ABSTRACT: This study aims to snapshot the existence of curriculum moderation in religious mentoring activities from three Indonesian universities. Religious mentoring is co-curricular activity in Islamic Religious Education (IRE) which focuses on character building. Anchored through a qualitative approach, we explored religious mentoring enacted in Universitas Pendidikan Indonesia (UPI), Universitas Negeri Malang (UM) and Universitas Tanjung Pura (UNTAN). Data was collected through interviews, documentation and observation. Findings suggest that religious mentoring is situated in a neutral position. Thus, religious moderatism is seen in mentoring activities through the formulation of a curriculum. Furthermore, religious mentoring activities can be used to understand Islamic moderatism. In this regard, the government needs to use regulations to direct mentoring activities to strengthen nationalism and moderatism.

Keywords: religious mentoring, religious moderation, curriculum moderation.

1 INTRODUCTION

Internalizing religious moderatism has been of great interest among researchers. It is due to the thoughts of intolerance and violence in the name of religion have an eternal nature (Abdallah 2019; Irham et al. 2020). In the context of Islam, historical viewpoints have uncovered the story of the upheaval of the *khawarij*'s group in the leadership of Ali bin Abi Talib until the emergence of the Islamic State of Iraq and Syria (ISIS) in the early 2000s (Mietzner & Muhtadi 2019; Sanders 2019). As one of the Muslim majority countries, Indonesia faces various social, religious, cultural, and political issues strictly related to religious intolerance and moderatism. This is proven by the recent attack on the Shiites group in Solo, Central Java, by a group of Muslims (Tempo 2020). In 2019, the Alvara Institute released their research related to the level of religious moderation in Indonesia. Religious views and rituals are the basis for formulating the religious typology of Muslims. Previous studies documented that that 56.7% of Muslims in Indonesia have a moderate view, 25.5% have a conservative view and 11.6% have an ultra-conservative view (Arifianto 2019; Rofiq et al. 2019).

A survey carried out by the Center for the Study of Islam and Society, State Islamic Universities in Syarif Hidayatullah, Jakarta in 2018, on 2,237 kindergarten to high school teachers identified that 43.5% or almost half of the respondents showed intolerant views toward non-Muslims (Laksana & Wood 2019; Sahrasad 2020). The radical thought within the survey was measured by some questions directed to the respondents. Another study also revealed that 27.59% of teachers prefer to advocate for people to join the war to create an Islamic country (Hashim & Langgulung 2008). Additionally, the Center for the Study of Islam and Society also discovered that kindergarten to secondary school students and university students are likely to be exposed to intolerance and radical Islamic teachings.

---

*Corresponding author: munjin.nasih.fs@um.ac.id

DOI 10.1201/9781003189206-3

University is an institution designed to establish a set of beliefs, values, and norms between generations (Hashim & Langgulung 2008). It is also expected to guide students into having a virtuous character with democracy and civilized values (Khisbiyah 2002). However, the recent intolerance acts within universities have called these ideal objectives into question. These agents propagate radical ideologies among the students. Consequently, universities currently fall under the scrutiny of being a place of radical transition, promotion, infiltration, birth, and growth (Abubakar & Hemay 2020; Fuad & Susilo 2019; Nur et al. 2020).

Some of the radicalism perpetrators are even found in public universities (Abubakar & Hemay 2020). In response to the aforementioned phenomena, the government has enacted several policies. One of these is described in Government Regulation No. 2, year 2017 on community organization, which covers the prohibition and disintegration of radical groups. Besides, the Ministry of Research, Technology, and Education Regulation No. 55 of the year 2018 on the Nation Ideology Construction within students' activity has also been legislated.

In addition to these policies, nationality and religiosity reinforcement has manifested in some universities' courses, such as Islamic religious education, citizenship education and Pancasila and character education courses. These courses are expected to enhance religious moderation in universities. Geared by such notions, the present study looked at religious mentoring activities enacted in three Indonesian universities (UM, UPI, UNTAN). Specifically, an investigation of the religious activities' curriculum formulation is the focus of this study. This study was conducted in three great Indonesian universities with a religious mentoring programs, namely UM, UPI, and UNTAN. These three universities are presumed to represent three mentoring programs with excellent administration and governance.

2 METHOD

This study employed a qualitative method carried out in three Indonesian universities, namely, UM, UPI, and UNTAN, from May to September 2020. Data was collected through interviews, document analysis, and observation. The data were then analyzed following Miles' (2014) qualitative analysis framework, which involves data condensation, data display, and inference. Furthermore, the data was analyzed using comparative analysis to identify religious moderation during the universities' mentoring activities.

3 RESULTS AND DISCUSSION

3.1 *Mentoring in UPI*

The program's curriculum is formulated based on the Islamic education curriculum at the university. This curriculum consists of Islam as *rahmatan lil âlamîn* (grace for the world); humans and religion; Islam comprehension methodology; *ijtihad* (establishment of Islamic law from al-Qur'an and Hadiths as the primary and secondary source, respectively); the development of faith and *taqwa* (piety); the teaching of worship and morals; halal and haram in Islam; marriage and family education; wealth management and usage; Madhhab (a school of thought within Islamic jurisprudence) and the Islamic school of thought; *amar ma'ruf nahu munkar* (acts commanded and forbade by God); da'wah; jihad; and contemporary issues (Purwanto & Fauzi 2019).

The information on mentoring activities in UPI was gathered from journals, student organization websites, and related literature (Kokasih et al. 2009).

3.1.1 *Duha lecture*

This lecture is held in the campus mosque with Islamic materials, with a 20 minute lecture and 40 minutes discussion or Q&A. The discussion session is longer to keep students engaged. The lecture is held from 08.00 to 10.00 a.m. The participants are asked to summarize the lecture once the activity is accomplished (Sinta et al. 2019).

3.1.2 *Mentoring*

This is a coaching model carried out in small groups lead by a tutor. The tutors are UPI students selected by the lecturer. The tutors are assigned to (1) deliver information to the participants, including the participants' tasks and stipulation; (2) conduct the learning and deliver the materials according to the syllabus; (3) review the materials; and (4) become facilitators during the discussion. This program is carried out in 12 weeks from Asr to Magrib prayer periods.

3.1.3 *Cadre formation, Bina Kader (Binder)*

This program is prepared for some elected students from each class. The participants of this program gain broader coaching. These participants are expected to carry positive changes in their classes.

3.1.4 *Integrated program*

This program is an additional optional program that facilitates students' interest and talent development, such as *tahsin, tahfidz*, Arabic, health, and authorship. The program is conducted together with related students' activity units.

3.2 Mentoring in UM

This university's mentoring curriculum consists of Islamic course material, Al-Qur'an recitation guidance, and worship guidance.

3.2.1 *General course material*

The selected general course materials are contextual and appropriate for teenagers. The materials are different from the ones discussed in the Islamic education course. Those materials include (1) positioning the heart to attain blessed knowledge; (2) teaching morals for students; (3) teaching unity of faith, sharia, and morals; (4) teaching patience and gratitude as the cure for confusion; (5) revealing the afterlife secret: the heaven and hell situations; (6) teaching various types of Islam in different parts of the world; (7) teaching language politeness in Islam; (8) teaching adolescent social ethics in Islam; (9) teaching the truth within the Al-Qur'an; (10) teaching the ideal Muslim student profile; (11) teaching the secrets of Muhammad's successful business; (12) teaching revitalization of the *Ikhtilaf al-fiqh* (disagreement of jurists) in responding to differences; (13) teaching repentance as a way to gain divine contentment; and (14) teaching *Syubban al-Yaum Rijal al-Ghad*: A youth today, a leader tomorrow (*Tafaqquh fi Dinil Islam's* schedule, odd semester 2019/2020).

From those 14 lecture topics, two topics (cases 6 and 12) lead to religious moderation reinforcement. The sixth topic promotes religious moderation since it presents information about the different Islam implementation worldwide. Thus, it confirms that differences in Islam are uncontested. Meanwhile, the 12th topic explicitly campaigns appreciation toward differences in Islamic law.

3.2.2 *Al-Qur'an recitation and worship guidance materials*

The materials in this activity are developed by a special team from the Al-Qur'an Study Club with instruction from Islamic education lecturers. The materials are published in a text book entitled "Pedoman Praktis Bimbingan Baca al-Qur'an (BBQ) Dilengkapi dengan Bacaan-bacaan di dalam Shalat" (Practical Guidelines for Reading the Al-Qur'an (BBQ) Guidance Equipped with Readings in Prayer). At the same time, the worship guidance materials are arranged by one of the Islamic education lecturers in a textbook, entitled "Fiqh Praktis. Panduan Mudah dan Lengkap memahami Tata Cara Bersesuci, Sholat, puasa, dan Zakat Fitrah" (Practical Fiqh. The Easy and Complete Guide to Understanding the Procedures for Purification, Prayer, Fasting, and Zakat Fitrah).

The religious moderation materials are identified in the last part of Al-Qur'an recitation guidance book. This part consists of prayer readings dominated by Syafi'iyah Madhhab and readings based on Maliki Madhhab (Tim Penyusun 2019). There are three types of activities covered in *the Tafaqquh fi Dinil Islam (TDI) program:* a general lecture, Al-Qur'an recitation guidance, and worship guidance.

Due to location constraints, the program was held on Saturdays, and the participants were divided into two groups. The odd Saturdays for group A and the even Saturdays for Group B. In contrast, the Al-Qur'an recitation and worship guidance programs are held in the working days, depending on the students' and the mentor's free time.

3.2.3 *General lecture*

This lecture consists of the religious speech delivered by Islamic education lecturers from Universitas Negeri Malang or other universities. The lectures are composed to be as enjoyable as possible to attract participants' enthusiasm. It is held for 60 minutes, from 06.30 to 07.30 a.m. To enhance the attractiveness and the participant's focus, the lectures use some learning media such as LCD projector, sound system, relevant movies and religious music. The lecture is followed by a discussion section, from 07.30 to 08.00 a.m. Besides, the students are assigned to write a summary to help them focus on the lecture. They have to write a resume on a book provided by the committee. To ensure the validity of the summary, the participants are also asked to get a stamp on their resume from the committee.

3.2.4 *Al-Qur'an recitation guidance*

This program is held from 08.00 to 09.00 a.m. The participants of this program are divided into groups based on the Al-Qur'an recitation ability. Each group consists of 12 to 15 students with a mentor. The materials involve Al-Qur'an recitation and *tajweed*. During the activity, the mentors check the participants' attendance and sign their books as proof of the participant's attendance.

3.2.5 *Worship guidance*

This activity is conducted from 09.00 to 09.30 a.m. The participants are divided into groups with one mentor in each group responsible for the activity in each group, similar to the Al-Qur'an recitation program. This program's activities are concentrated on guidance for wudu', prayer readings, and praying procedures during a journey. The participants' daily worships are also monitored, including both obligatory and Sunnah worship, such as compulsory prayer, Sunnah prayer, Sunnah fasting, and so forth. During the activities, the mentor also checks the participants' attendance.

3.3 Mentoring in UNTAN

UNTAN's character education *(Pendikar)* identity and special features are represented by three keywords: brotherhood, involving the several academic disciplines, and tolerance. Annually, more than 7000 freshmen of Universitas Tanjungpura are tied into a small family of *Pendikar*. These students come from various disciplines and try to know each other and learn and comprehend each other's fields. Further, they are divided based on their religious groups such as Islamic, Catholic, Christian, Buddhist, Hindu, and Confucian to learn their religions. Each weekend, they have an interfaith discussion intending to grow tolerance between religious believers and realize harmony, cooperation, and respect among them, following article 5 paragraph 4 of Government Regulation No. 55 of the year 2007 on religious and spiritual education. The character education program *(Pendikar)* in UNTAN is enacted in so the freshmen learn about God, as well as strengthen the friendships among them and the faculty members. The implemented method asks students to read the translation of their scripture and perform prayers every day while taking some time to communicate with other students, at least once a week. The methods are evaluated and improved by the tutors every Friday afternoon. Therefore, an improvement in human relations with God, as well as with other humans, is attained, along with the expected virtuous behaviors.

The implementation *Pendikar* in each religion had some similarities. They held regular group meetings every Friday from 01.00 to 03.00 p.m. They also had friendship visits every weekend and were obligated to learn their scriptures. In its implementation, the participants, tutors, and lecturers use Facebook to ease communication and coordination. This social media platform was chosen since it is easy to use and has many users.

4 CONCLUSION

The present study has captured that the religious mentoring activities enacted in three Indonesian universities (UPI, UM and UNTAN). These activities are carried out to nurture the Pancasila morals among the students. As a national concept and ideology, Pancasila has great value within Indonesia's social and cultural advancement. In other words, the activities in those universities ask students to memorize and learn this national concept and implement it in their daily lives. Such religious mentoring activities are also intended to internalize religious moderation and reinforce love for the nation among Indonesian university students.

REFERENCES

Abdallah, A. 2019. State, Religious Education, and Prevention of Violent Extremism in Southeast Asia. *Studia Islamika*, *26*(2), 407–415.

Abubakar, I., & Hemay, I. 2020. Pesantren Resilience: The Path to Prevent Radicalism and Violent Extremism. *Studia Islamika*, *27*(2).

Arifianto, A. R. 2019. Islamic campus preaching organizations in Indonesia: Promoters of moderation or radicalism? *Asian Security*, *15*(3), 323–342.

Fuad, A. J., & Susilo, S. 2019. Mainstreaming Of Islamic Moderation In Higher Education: The Radical Experience Conter. *Proceedings of Annual Conference for Muslim Scholars*, *3*, 467–483.

Hashim, C. N., & Langgulung, H. 2008. Islamic religious curriculum in Muslim countries: The experiences of Indonesia and Malaysia. *Bulletin of Education & Research*, *30*(1), 1–19.

Irham, I., Haq, S. Z., & Basith, Y. 2020. Deradicalising religious education: Teacher, curriculum and multiculturalism. *Epistemé: Jurnal Pengembangan Ilmu Keislaman*, *15*(1), 39–54.

Kokasih, A., Fahrudin, & Anwar, S. 2009. Pengembangan Model Pembelajaran PAI Melalui Pembinaan Keagamaan Berbasis Tutorial. *Jurnal Penelitian*, *9*(1).

Laksana, B. K., & Wood, B. E. 2019. Navigating religious diversity: Exploring young people's lived religious citizenship in Indonesia. *Journal of Youth Studies*, *22*(6), 807–823.

Mietzner, M., & Muhtadi, B. 2019. The Mobilization of Intolerance and its Trajectories: Indonesian Muslims' Views of Religious Minorities and Ethnic Chinese. *Contentious Belonging: The Place of Minorities in Indonesia*, 155–174.

Nur, I., Nawawie, A. H., Fajarwati, H., & Chusna, H. 2020. Embracing Radicalism and Extremism in Indonesia with the Beauty of Islam. *Asian Research Journal of Arts & Social Sciences*, 1–18.

Purwanto, Y., & Fauzi, R. 2019. *Internalisasi Nilai Moderasi Melalui Pendidikan Agama Islam Di Internalizing Moderation Value Through Islamic Religious Education*. *17*(2), 110–124.

Rofiq, A. C., Mujahidin, A., Choiri, M. M., & Wakhid, A. A. 2019. The Moderation of Islam In The Modern Islamic Boarding School of Gontor. *Analisis: Jurnal Studi Keislaman*, *19*(2), 1–24.

Sahrasad, H. 2020. *A'an Suryana. The State and Religious Violence in Indonesia: Minority Faiths and Vigilantism*. Center for Southeast Asian Studies, Kyoto University.

Sanders, C. E. 2019. *Comparing Religious Intolerance Across Religious Groups: An Analysis of Covariance Using Self-Report Survey Responses* [PhD Thesis]. Northcentral University.

Sinta, D., Syahidin, & Hermawan, W. 2019. Peran Tutorial Pai Dalam Menangkal Paham Radikal Keagamaan di Kampus UPI. *Tarbawy*, *6*(1), 1–18.

Community Empowerment through Research, Innovation and Open Access – Sayono et al (Eds)
© 2021 Copyright the Author(s), ISBN 978-1-032-03819-3

Improving teachers' teaching abilities in the era of the ASEAN Economic Community by the teacher's professional coaching

A. Imron, B.B. Wiyono, I. Gunawan, B.R. Saputra*, D.B. Perdana & S. Hadi
Universitas Negeri Malang, Indonesia

A. Abbas
International Islamic University Islamabad, Pakistan

ABSTRACT: The main concept of ASEAN Economic Community (AEC) is to make ASEAN a single market and production-based unification by 2015. As part of ASEAN, Indonesia has to integrate its economic and human resource capabilities in the AEC. It is necessary to identify the characteristics of the AEC, as well as prepare the students and the teachers to teach people who are ready to face the AEC era. To provide those qualified teachers, continuous professional coaching is necessary to improve their commitments and teaching capabilities. Therefore, research that produces a model of professional teacher coaching that ensures the realization of teachers' instructional skills for the AEC era is required. It is necessary to get guidance from previous studies that have done that in facing global challenges. This professional coaching aims to improve teachers' teaching skills through a variety of supervision techniques.

Keywords: Teacher's professional coaching, teacher professionalism, teaching abilities, economics community.

1   INTRODUCTION

The leaders of the governments of ASEAN countries in October 2014 created an agreement by signing the ASEAN Concord-2 in Bali. Among these agreements, the ASEAN Economic Community (AEC) was established as the realization of the ultimate goal of economic integration as outlined in the ASEAN Vision 2020, which is to create a stable, prosperous and highly competitive ASEAN economic region (Ministry of Trade 2013). In the ASEAN region, there will be the free trade of goods, services, investments, and capital flows (Mantra 2011). The AEC blueprint had been declared at the 13th ASEAN Summit in Singapore in 2007 (ASEAN Secretariat 2008).

Reviewing the student ability survey released by the Program for International Student Assessment (PISA) in 2019, Indonesia is ranked 72nd out of 77 countries (Viva 2019). On the other side, the news portal Tirto released the Education Index data based on Human Development Reports (2017). According to the data, Indonesia is placed seventh in ASEAN countries with a score of 0.622. This ranking is still considered lagging behind other ASEAN countries, such as Singapore (0.832), Malaysia (0.719), Brunei Darussalam (0.704), Thailand, and is just ahead of the Philippines (0.661). Ben-Yehuda (2015) stated that the teacher professional coaching program had a significant effect on the students' academic ability. Teacher professional coaching is a crucial aspect that should be developed sustainably. This development can be built through good relationships in training, self-reflection of each individual on their abilities, experiences on challenging new insights and experiments with new ideas (Patti et al. 2012). The optimization in every stage of teacher professional development should be integrated with the support of the school principals and supervisors.

---

*Corresponding author: bagusrachmad47@gmail.com

DOI 10.1201/9781003189206-4

The substance of the blueprint was the strategic schedule for realizing the 2015 AEC. The professionalism that has been developed this way is expected to impact the teacher's teaching ability in the learning process. According to the research conducted by Sudjana (2002), the professionalism of teachers has quite dominantly influenced the student's learning process by 76%. The teacher's professionalism includes 32.43% for teaching ability, 32.38% for mastery of the subject matter, and 8.60% for teacher's attitudes towards related subjects. Based on the background of the problem and the lines of thought described above, this paper further examines the appropriate teacher professional coaching to improve their teaching abilities in facing the era of the AEC, entitle "Teacher's Professional Development to Improve the Teaching Ability of Teachers in the Era of the AEC."

2 TEACHER'S PROFESSIONAL COACHING IN THE AEC ERA

Professional coaching has established ethics, competencies, proficiencies, and masteries that take the process of adult learning far beyond what was possible before with the help of particular super-visors and their advice on how to manage classrooms or teach lessons. The coaching emphasizes that the teacher's professionalism improvement should be the focus of the teacher's professionalism coaching as well as building their commitment through mentorship. Teacher professional coaching also covers training in the adaptation of the profession of teachers and managers in education systems. In addition, the purpose of the teacher's professional coaching is to improve the teacher's commitment with the influence of leadership and working conditions on teacher commitment as well as with the development and maintenance of high levels of commitment among teachers (Imron 2016).

Sobri (2013) explained that professional coaching for teachers is an attempt or activity carried out in a structured manner to obtain optimal results for teachers in self-development and learning. Teacher's professional coaching, according to the Ministry of Education and Culture, is a series of activities that provide assistance to teachers, especially in the form of professional service assistance, which is carried out by school principals, school owners, supervisors, and others to improve the teaching and learning process and outcome.

In creating high-quality lesson planning, the teacher's ability is required in order to design a learning process that refers to qualitative and quantitative improvisations. Usman (2009) describes that ability or competence should be the primary consideration in rationalizing a predetermined goal according to actual conditions. Thus, the teacher's teaching ability is a self-reflection of their capabilities, encouraging students to learn and develop continuously (Sauri 2010). Commitment is an essential element of successful teaching; committed teachers are concerned with their students' development and profoundly struggle with how to keep students learning. Increasing teacher commitment in teaching will have a positive impact on student learning outcomes. The teacher will really focus and pay attention to students, not just doing work but truly committed to helping students find their potential. Teacher commitment has been identified as one of the most critical factors for the future success of education and schools. Teachers committed to the teaching profession are thought to be more satisfied with the job and are likely to identify themselves as teachers. The level of teacher commitment is considered a critical factor in the success of the current educational reform agenda as it heavily influences teachers' willingness to engage in cooperative, reflective and critical practice (Imron 2016).

3 TEACHING ABILITY IN THE AEC ERA

AEC is one of the pillars of the long-term plan launched by ASEAN. The purpose of the AEC itself is to manifest an integrated system in the economic field or a single market implemented by the countries in it. The improvements in the teacher's ability have a significant effect on producing better students. The demands for highly competitive professionals in the AEC era are the key to be able to compete in the AEC single market flow. In the AEC itself, there are challenges in employment, one of which is to present competitions in terms of education, productivity, skills, and abilities. The teacher's ability will affect the learning outcomes that the students will achieve. The essence of being an effective teacher lies in knowing what to do to foster pupils' learning and being able to do it (Kyriacou 2007). Rubio (2009) divided teachers' teaching abilities into two aspects: personal abilities and professional abilities. The teacher's professional skills are content knowledge, good planning, clear goals and communication, good classroom management and organization, and consistently high and realistic expectations with the students are essential factors to be effective teachers (Rubio 2009). The teacher's personal skills are caring, knowing the students individually, fostering teacher-student relationships, and creating an effective classroom environment (Rubio 2009).

In order to be prepared to manage AEC with its competitive trade traits, services, and professionals, a professional teacher with better abilities is needed. Therefore, teachers' teaching ability needs to be optimized to educate students who are already prepared to approach the AEC era. A high level of teaching ability is necessary, but it is still not sufficient to face the AEC era if it is not being complemented with a high commitment to educating the students. That is why the teacher's commitment and teaching ability are two inseparable sides of the same coin (Imron 2016). Optimization of each side is needed through professional coaching, both by school principals and school supervisors. The knowledge, skills, and commitment of teachers, as well as the quality of school leadership, are the essential factors in achieving high-quality educational outcomes (European Commission 2013).

4 TEACHER'S PROFESSIONAL COACHING TO OPTIMIZE THE TEACHER'S TEACHING ABILITY IN THE AEC ERA

In order to provide proper teacher professional coaching, Dreyfus and Rabinow (1986) recommend standing on the teacher's level of development on their teaching maturity and abilities. The first level is the novice level, when the teacher presumes that personal practical experience is more valuable than verbal information. Teachers at this level are taught the meaning of certain terms and concepts, school rules, and the situation's objectives and characteristics. Second, the advanced level starts from influencing their behavior in a meaningful way. At this level, teachers do not feel a sense of independence/autonomy concerning their work. Teachers still feel completely irresponsible for their actions.

The third level is the competent level, or when the teacher has moved and has enough experience and motivation to succeed. Fourth, the proficient level is when the teacher begins to recognize formulas and equations holistically. Fifth, the expert level is when the teacher has demonstrated performance and decision-making intuition. They present it in a distinctive and qualitative way to other teachers. Kasule et al. (2014), in their research, stated that the teacher's professional coaching program should involve five domains/areas of competences that have to be achieved by a teacher that acts as: (1) an innovator; (2) a community science developer; (3) a network maker; (4) an educational designer; and (5) an entrepreneur. Kabilan (2004) shows that there are five aspects of teacher competence that are useful for improving the teacher's competence, namely: (1) motivation; (2) knowledge and skills; (3) independent learning; (4) interaction competencies; and (5) awareness of technology.

Mevarech (1995) describes several steps in the process of teacher's professional coaching, namely: (1) survival; (2) exploration and bridging; (3) adaptation; (4) conceptual changes; (5) discovery; and (6) experiments. Contingency theories recommend a model of teacher's professional coaching based on the level of maturity (LoM), level of responsibility (LoR), and concern of teacher (CoT). Teachers with a mature LoM, independent LoR, and professional CoT are recommended to use a nondirective model. Meanwhile, teachers with immature LoM, dependent LoR, and self-CoT are advised to use the directive model. On the other side, teachers with growing LoM, autonomous LoR, and CoT students are recommended to use the directive model.

5 CONCLUSION

In confronting the era of the ASEAN Economic Community (AEC), improvements in the educational sector are very fundamental. These improvements aim to produce teachers who are qualified to be educators who are competitive with foreign teachers. In addition, teachers with qualified teaching qualities will produce high-achieving and talented students as well. Teacher professional coaching is conducted to support the teachers in improving their teaching abilities and awareness of their responsibilities as educators. Teacher's professional coaching is organized systematically by the principal, school supervisor, or the education office. Some issues that must be considered in conducting teacher's professional coaching programs are the integration of a set of structures, concepts, and competencies in value education that the teachers have to be master. The professional coaching activities that are being carried out should be cooperative, collaborative and constructive based on the teacher's empirical experience in learning. Thus, the deficiencies in learning activities can be evaluated by providing solutions and innovations to creating an optimized learning objective according to the circumstances.

REFERENCES

ASEAN Secretariat. 2008. ASEAN Economic Community Blueprint. Jakarta: Public Affairs Office, ASEAN Secretariat.
Ben-Yehuda, M. 2015. *The Route to Success – Personal Academic, Coaching Program.* International Conference Education, Reflection, Development (ERD 2015). Social and Behavioral Sciences.
Dreyfus, H, and P Rabinow. 1986. What is Maturity? Foucault and Habermas on What is Enlightenment? In Foucault, M., and Hoy, DC Oxford: Blackwell: Foucault: A Critical Reader.
European Commission. 2013. Supporting Teacher Competence Development: For Better Learning Outcomes. Brussels: European Commission.
Imron, Ali. 2016. Strengthening School Quality Management in the MEA Era. Paper presented in the Seminar on Strengthening Quality Management and Educational Leadership in the MEA Era and Globalization, Postgraduate Program. Malang: State University of Malang.
Kabilan, MK 2004. Online Professional Development: A Literature Analysis of Teacher Competency. Journal of Computing in Teacher Education.
Kasule, GW, R Wesselink, O Noorozi, and M Mulder. 2014. The Current Status of Teaching Staff Innovation Competence in Ugandan Universities: Perceptions of Managers, Teachers, and Students. Journal of Higher Education Policy and Management.
Kyriacou, C. 2007. Essential Teaching Skills. London: Nelson Thornes, Ltd.
Mantra, D. 2011. Hegemony and the Discourse of Neoliberalism: Tracing Indonesia's Steps towards the 2015 ASEAN Economic Community. Jakarta: Mantra Press.
Mevarech, ZR 1995. Teachers Paths on the way to and from the Professional Development Forum. New York: Teachers College Press.
Ministry of Trade. 2013. Towards the 2015 ASEAN Economic Community. Jakarta: Ministry of Trade of the Republic of Indonesia.
Patti, J, AA Holzer, R Stern, and MA Brackett. 2012. Personal, Professional Coaching: Transforming Professional Development for Teachers and Administrative Leaders. *Journal of Leadership Education* 11(1).
Rubio, CM 2009. Effective Teachers: Professional and Personal Skills. Ensayos, Revista de la Facultad de Educación de Albacete.
Sauri, Sofyan. 2010. *Building national character through teacher professionalism development based on value education.* Journal of Character Education 2 (2): 1–15.
Sudjana, Nana. 2002. Assessment of Teaching and Learning Process Results. Bandung: Youth Rosdakarya.
Sugiyono. 2016. Metode Penelitian Kuantitatif Kualitataif dan Kombinasi (Mixed Methods). *Journal of Chemical Information and Modeling.*
Sobri, Ahmad Yusuf. 2013. *Fostering Teacher Professionalism in Improving Learning Quality.* Journal of Educational Management 24 (1): 9–20.
Tirto. 2019. *Indonesia's Education Index is Low, Competitiveness is Weak.* tirto.id. Accessed 12 October 2020. https://tirto.id/indeks-pend Pendidikan-indonesia-rendah-daya-saing-pun-lemah-dnvR.
Uno, Hamzah B. 2008. *Learning Model.* Jakarta: PT Bumi Aksara.
Viva. 2019. *World Education Survey, Indonesia Ranks 72 out of 77 Countries.* December 5, 2019. https://www.viva.co.id/arsip/1249962-survei-pend Pendidikan-dunia-indonesia-per Rank-72-dari-77-negara.

Community Empowerment through Research, Innovation and Open Access – Sayono et al (Eds)
© 2021 Copyright the Author(s), ISBN 978-1-032-03819-3

Exploring the relationship among polarity, subjectivity, and clusters characteristics of visitor review on tourist destination in Malang, Indonesia

A. Larasati*, M. Farhan, P. Rahmawati, J. Sayono & A. Purnomo
Universitas Negeri Malang, Indonesia

E. Mohamad
Universiti Teknikal Malaysia Melaka, Malaysia

ABSTRACT: The development of information technology has made it easier for various people activities. One of them is the development of information technology in the tourism sector, where information and access to deliver opinions or reviews on tourist attractions are available on various websites and social media. This study aims to explore the relationships among polarity, subjectivity, and clusters characteristics of online tourist reviews on different tourist objects in the Malang Regency. Online-review data was retrieved using the web content mining technique. The study applies sentiment analysis and k-mean clustering to analyze retrieved online-review data. The clustering process was performed to classify existing reviews based on their respective characteristics using the k-means clustering algorithm. The results show a significant difference between clusters and sentiment of the visitor reviews on each tourist object. The sentiment analysis results also indicate differences in the interpretation of polarity and subjectivity in each cluster at the same objects. The results of each tourist object's clusters and sentiment show that the review of the services and objects offered tends to be a positive sentiment. On the other hand, the reviews for the facilities offered, the visitors' reviews tend to be neutral. For the negative sentiment, it mostly occurs about the price as well as the location of the tourist objects.

1 INTRODUCTION

The sentiment is one of the most important things for any company or business. Sentiment analysis can help see feelings, needs, and patterns of customer responses to the product or service being offered (Syakuyaqubr et al. 2018). In sentiment analysis, two parameters are used as measurement benchmarks, namely the polarity and subjectivity values generated from online reviews delivered by customers. Polarity shows whether the reviews contain positive, neutral, or negative feelings. Meanwhile, subjectivity points out whether the responses can be classified as visitors' opinions or not. In terms of polarity, the higher the value generated increases the likelihood of the positive responses, while higher subjectivity indicates that the review given is an opinion. Various techniques can be used to perform sentiment analysis, for example, using machine learning. Machine learning can determine the pattern of comments given by visitors to help managers understand the needs of visitors and their feelings (Syakur et al. 2018).

The clustering process aims to see the grouping of reviews in each tourist destination using K-means. It is the most commonly used algorithm in clustering because of its simplicity of use. The advantage of the K-means method is its effective use on large amounts of data. However, the K-means method has limitations in determining the best number of clusters. This limitation is in

---

*Corresponding author: aisyah.larasati.ft@um.ac.id

DOI 10.1201/9781003189206-5

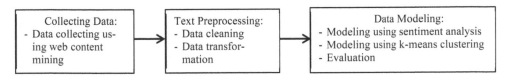

Figure 1.   Research methods (collecting data, text processing, and data modeling).

the form of the final cluster result, which is very dependent on determining the center of the cluster (centroid) at the beginning of the process. Thus this algorithm is considered to have high sensitivity because the determination of the centroid's starting point has an impact on the final cluster results (Syakur et al. 2018).

This study aims to explore the relationship between clustering and sentiment (polarity and subjectivity) from online reviews given by visitors of tourist destinations in the Malang Regency. The polarity and subjectivity results obtained were linked to the characteristics of the cluster formed. Online reviews were obtained by implementing web content mining at each tourist destination. Web content mining works by indexing information in the form of text, audio, images, HTML, XML, which consists of structured, semi-structured, and non-structured data types (Mebrahtu & Srinivasulu 2017). The advantage of web content mining is the capability to change unstructured data into structured data easily understood by humans.

Sentiment analysis is a technique used to identify and classify the polarity and subjectivity of a text (Ankit & Saleena 2018). The process of sentiment analysis uses Natural Language Processing (NLP) works by assessing the emotional tendency of a word or sentence (Drus & Khalid 2019). Polarity analysis is useful to determine whether the opinion falls into the positive, neutral, or negative category. In polarity analysis, the words in the lexicon are weighted with a score ranging from -1 for negative sentiment, 0 representing neutral sentiment, and +1 for the positive sentiment (Drus & Khalid 2019). Within the scope of sentiment analysis, there are stages of subjectivity analysis. Subjectivity analysis is the process of classifying an opinion as a subjective opinion or an objective word of fact. In the sentiment analysis, words are scored with a 0 to 1 score; a score close to 1 means the text is subjective, while a score close to 0 represents objective opinion (Yaqub et al. 2018). Sentiment analysis is useful and widely applied in various fields such as e-commerce, tourism, health, manufacturing, and various other industries (Irawan et al. 2019). The existence of sentiment analysis can be a reference for business managers in making continuous improvements to a business object's facilities and services.

2   METHODS

The method of this study is shown in Figure 1. Data was extracted from online visitor reviews on five tourist objects in Malang Regency available on various websites. The tourist objects were *Masjid Turen, Bedengan, Pantai Tiga Warna, Kebun Teh Wonosari, and Lembah Indah.* The total number of reviews obtained was 4269 data, but only 2170 data were analyzed after the text preprocessing phase.

The process of collecting visitor reviews on each tourist attraction used web content mining technology. Web content mining is a part of the web mining field that can be used for collecting unstructured data sourced from websites by accessing the specific HTML elements of the data (Slamet et al. 2018). The web content mining process was started with selecting web page URL, namely websites of tourist attractions in Malang, which have visitor reviews. Web content mining can process data in multiple web pages at one time, thereby increasing time efficiency while retrieving the data. The retrieved data was used as data input in undertaking the analysis process by applying data mining techniques, including text mining and sentiment analysis, to attain new insight as a fundamental for better decision making (Hassani et al. 2020).

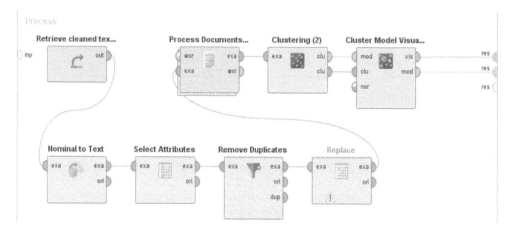

Figure 2.   Text preprocessing and K-means clustering in Rapid Miner.

After collecting data, the text preprocessing was carried out. This method transforms text data into a numerical format based on the vector dimensions that are formed in each composing word. After all the text preprocessing steps had been completed, the files were saved in excel form. Then the process of Sentiment Analysis and K-means clustering was continued by utilizing Rapid Miner.

The outcomes of K-means were further evaluated by considering the polarity and subjectivity values of Sentiment Analysis to discover and determine whether there was a correlation among polarity, subjectivity and cluster characteristics. The detailed steps from text preprocessing to K-means algorithm implementation in Rapid Miner are shown in Figure 2.

3   RESULTS AND DISCUSSION

The following is the result of data processing with 2170 reviews extracted from five tourist objects within the Malang Regency. In the clustering process, the number of clusters was determined by 5 clusters. The evaluation of clustering results was carried out by calculating the Davies-Bouldin Index (DBI) score. The best cluster results are shown in the smaller DBI score. The wordlists are displayed in the table covering the top 5 rankings on visitors' reviews of tourist objects.

3.1   *Masjid Turen*

Table 1 denotes the characteristics and sentiment in each cluster with 532 reviews obtained from the websites related to the tourist object of *Masjid Turen*. The value of the Davies Bouldin Index for the cluster is 2.69. Referring to polarity for discovering sentiment categories and subjectivity to ensure either fact or opinion in each review, visitors tend to give positive reviews based on their views and experiences, with the overall percentage being 87.4%. In contrast, the rest of the visitors tend to deliver neutral reviews. There are no negative sentiments given by visitors.

3.2   *Bedengan*

Table 2 denotes the characteristics and sentiment in each cluster with 389 reviews obtained from the websites.

The value of the Davies Bouldin Index for the cluster shown in Table 2 is 4.76. Referring to polarity for discovering sentiment categories and subjectivity for ensuring either fact or opinion in each review, visitors tend to give positive reviews based on their opinions and experiences. The overall percentage is 61.95%. Besides, the percentage of negative sentiments and neutral sentiments is 22.62% and 15.42%, respectively.

Table 1. Cluster characteristics of *Masjid Turen*.

No	Wordlist	Total Occurrences	Total items in each cluster	Characteristics	Sentiment
	Wordlist Occurrences		Stemming		
1	Mosque	887	67	Tourism, religion, floor	Neutral
2	Building	202	183	Build, School, Tourist	Positive
3	Beautiful	132	84	Board. School, tourism	Positive
4	Cool	108	118	Cool, place, good	Positive
5	Unique	90	80	Unique, religious, tourism	Positive

Table 2. Cluster characteristics of *Bedengan*.

No	Wordlist	Total Occurrences	Total items in each cluster	Characteristics	Sentiment
	Wordlist Occurrences		Stemming		
1	Place	228	88	Car, malang, unfortune	Negative
2	Cool	152	113	Fresh, recommend, nature	Positive
3	Natural	102	59	Picnic, Suitable, Family	Positive
4	Beautiful	81	69	Comfort, Relax, beauty	Positive
5	Good	68	60	Tree, flow, pine	Neutral

Table 3. Cluster characteristics of *Pantai Tiga Warna*.

No	Wordlist	Total Occurrences	Total items in each cluster	Characteristics	Sentiment
	Wordlist Occurrences		Stemming		
1	Beach	905	166	Great, fun	Positive
2	Beautiful	165	19	Difficult, access, look	Negative
3	Clean	143	66	Beauty paid, suitable	Positive
4	Good	127	66	Number, day, people	Neutral
5	Trash	88	75	Ojek, taxi, clungup	Neutral

3.3 *Pantai Tiga Warna*

Table 3 denotes the characteristics and sentiment in each cluster with 392 reviews obtained from the websites. The value of the Davies Bouldin Index for clustering is 4.8. Referring to polarity for discovering sentiment categories and subjectivity to ensure either fact or opinion in each review, visitors tend to give positive reviews based on their opinions and experiences, with the overall percentage being 59.18%. In addition, the rate of negative sentiments and neutral sentiments are 4.84% and 35.96%, respectively.

3.4 *Kebun Wonosari*

Table 4 denotes the characteristics and sentiment in each cluster with 575 reviews obtained from the websites. The value of the Davies Bouldin Index for the cluster is 3.08. Referring to polarity for discovering sentiment categories and subjectivity for ensuring either fact or opinion in each review, visitors tend to give positive reviews based on their opinions and experiences. The overall positive review percentage is 75,13 %, while the rest is neutral reviews.

Table 4. Cluster characteristics of *Kebun Wonosari*.

No	Wordlist	Total Occurrences	Total items in each cluster	Characteristics	Sentiment
	Wordlist Occurrences		Stemming		
1	Tea	665	116	Air, comfort, suitable	Positive
2	Garden	301	140	Pool, swim, children	Positive
3	Good	159	103	Good, view, nice	Positive
4	Beautiful	130	143	Wonosari, tea, gardern	Neutral
5	Fresh	67	73	Spot, photo, tourist	Positive

Table 5. Cluster characteristics of *Lembah Indah*.

No	Wordlist	Total Occurrences	Total items in each cluster	Characteristics	Sentiment
	Wordlist Occurrences		Stemming		
1	Cool	135	76	Air, family, cool	Positive
2	Good	98	42	Ticket, prie, facility	Neutral
3	Beautiful	90	74	Tourist, Malang, facility	Positive
4	Nice	45	39	Photo, spot, good	Positive
5	Natural	35	51	Nice, view, place	Positive

3.5 *Lembah Indah*

Table 5 denotes the characteristics and sentiment in each cluster with 282 reviews obtained from the websites. The value of the Davies Bouldin Index for the clusters is 2.683. Referring to polarity for discovering sentiment categories and subjectivity for ensuring either fact or opinion in each review, visitors tend to give positive reviews based on their opinions and experiences, with the overall percentage are 85.1%, while the rest tend to deliver neutral reviews.

Based on the results shown in Tables 1–5, each tourist attraction has different cluster characteristics. These characters may reflect the visitors' tendency. The expressions delivered by visitors can be in the form of evaluations or their experiences and descriptions related to the tourist objects visited. The cluster characteristics depend on the wordlist and its occurrences, which are derived from the web scrapping process. Web scraping is the process of extracting, collecting, and storing data sourced from the internet into a database (Dewi et al. 2019). The web scraping process has various advantages, such as it is easy and does not require much data processing time. Besides, the web scraping process can also convert unstructured data into structured forms such as spreadsheets and CSV. The existence of web scraping helps in various aspects, for example, identify web changes, competitor analysis, stock prices, market data, weather data, and so forth (Saurkar & Gode 2018). Therefore, this process is considered to contribute to business processes when making optimal decisions based on data, including extracting cluster characteristics.

4 CONCLUSION

The data processing results on five tourist objects reveal a significant difference between clusters and sentiment of the visitor reviews on each tourist object. The DBI calculation results show that the clusters resulted for *Bedengan* and *Pantai Tiga Warna* are worse than the others. The characteristics of the words that appear in each cluster also have significant differences between each tourist object. The sentiment analysis conducted shows differences in the interpretation of polarity and subjectivity in each cluster at the same objects. Thus, to improve tourist visits, tourist management may focus on clusters that deliver negative sentiments while maintaining good service for clusters that deliver positive reviews.

REFERENCES

Ankit, Saleena, N., 2018. An Ensemble Classification System for Twitter Sentiment Analysis. *Procedia Computer Science* 132, 937–946. https://doi.org/10.1016/j.procs.2018.05.109

Dewi, L.C., Meiliana, Chandra, A., 2019. Social media web scraping using social media developers API and regex. *Procedia Computer Science* 157, 444–449. https://doi.org/10.1016/j.procs.2019.08.237

Drus, Z., Khalid, H., 2019. Sentiment analysis in social media and its application: Systematic literature review. *Procedia Computer Science* 161, 707–714. https://doi.org/10.1016/j.procs.2019.11.174

Hassani, H., Beneki, C., Unger, S., Mazinani, M.T., Yeganegi, M.R., 2020. Text mining in big data analytics. *Big Data and Cognitive Computing* 4, 1–34. https://doi.org/10.3390/bdcc4010001

Irawan, H., Akmalia, G., Masrury, R.A., 2019. Mining tourist's perception toward Indonesia tourism destination using sentiment analysis and topic modelling. *ACM International Conference Proceeding Series* 7–12. https://doi.org/10.1145/3361821.3361829

Mebrahtu, A., Srinivasulu, B., 2017. Web Content Mining Techniques and Tools. *International Journal of Computer Science and Mobile Computing* 6, 49–55.

Saurkar, A. V, Gode, S.A., 2018. An Overview On Web Scraping Techniques And Tools. *International Journal on Future Revolution in Computer Science & Communication Engineering* 363–367.

Slamet, C., Andrian, R., Maylawati, D.S., Suhendar, Darmalaksana, W., Ramdhani, M.A., 2018. Web Scraping and Naïve Bayes Classification for Job Search Engine. *IOP Conference Series: Materials Science and Engineering* 288. https://doi.org/10.1088/1757-899X/288/1/012038

Syakur, M.A., Khotimah, B.K., Rochman, E.M.S., Satoto, B.D., 2018. Integration K-Means Clustering Method and Elbow Method for Identification of the Best Customer Profile Cluster. *IOP Conference Series: Materials Science and Engineering* 336. https://doi.org/10.1088/1757-899X/336/1/012017

Yaqub, U., Sharma, N., Pabreja, R., Chun, S.A., Atluri, V., Vaidya, J., 2018. Analysis and visualization of subjectivity and polarity of twitter location data. *ACM International Conference Proceeding Series*. https://doi.org/10.1145/3209281.3209313

Community Empowerment through Research, Innovation and Open Access – Sayono et al (Eds)
© 2021 Copyright the Author(s), ISBN 978-1-032-03819-3

Stigma and dilemma: An ethnographic review on the medical personnel of the COVID-19 referenced hospital

D.W. Apriadi*, D. Mawarni & M. Saputra
Universitas Negeri Malang, Malang, Indonesia

ABSTRACT: The high number of COVID-19 cases in Indonesia has caused anxiety among the Indonesian people and has created a negative stigma attached to medical personnel because they are considered vulnerable to being exposed to COVID-19. The purposes of this research were (1) to find out the negative stigma attached to medical personnel, both from the family and society in the midst of the COVID-19 pandemic, (2) to explore the impact of negative stigma in the midst of the COVID-19 pandemic, and (3) to investigate the solutions applied to prevent and overcome the negative stigma amid the COVID-19 pandemic. This research employed a qualitative research approach, and the data was gathered using in-depth interviews via video calls and/or by telephone. The results of this study indicate that the negative stigma received by medical personnel comes from family, friends, neighbors, and even immediate family members. The attitudes shown are also of various kinds, such as discrimination, exclusion, bullying, and even expulsion.

Keywords: Stigma, Medical Personnel, COVID-19 Pandemic

1   INTRODUCTION

The beginning of the emergence of the first COVID-19 case in the City of Surabaya was March 17, 2020. The Ministry of Health of the Republic of Indonesia then issued a decree on the determination of Large-Scale Social Restrictions (PSBB) number HK.01. 07/MENKES/264/2020 on April 21, 2020. The East Java Provincial Government issued a Governor of East Java Decree number 188/202/KPTS/013/2020, stipulating the Enforcement of Large-Scale Social Restrictions (PSBB) in handling Covid-19 in the region. Surabaya City, Sidoarjo Regency, and Gresik Regency, for 14 days, from 28 April 2020 to 11 May 2020, issued the Governor of East Java Decree number 188/219/KPTS/013/2020 concerning the extension of PSBB's jurisdiction in handling COVID-19 in the City of Surabaya, Sidoarjo Regency, and Gresik Regency for 14 days starting from May 12, 2020, to May 25, 2020.

After the end of the PSBB, the mayor of Surabaya issued a regulation of the mayor of Surabaya number 28 of 2020 concerning Guidelines for the New Normal Order in the Conditions of the COVID-19 Pandemic in the City of Surabaya on June 10, 2020 as a guideline for implementing the new normal order in the conditions of the COVID-19 pandemic which includes changes in the culture of life communities which would be more productive in the COVID-19 pandemic situation by implementing a Clean and Healthy Lifestyle (PHBS) and health protocols which were expected to reduce the risk and impact of COVID-19. The increase in the number of COVID-19 patients raised anxiety in the community, which also raised various negative stigmas, especially for medical personnel who work in hospitals and are vulnerable to being exposed to COVID-19. In the United States study, nearly half were found to be feeling restless, according to a survey from the American Psychiatric Association (Varshney 2020). The psychological impact experienced by health workers is in the forms of stress and anxiety due to the social stigma that exists in society.

---

*Corresponding author: deny.apriyadi.fis@um.ac.id

DOI 10.1201/9781003189206-6

The stigmatization of health workers adds to the physical and mental fatigue that health workers bear during their working hours. Doctors or paramedics trying to save lives during these difficult times are refused habitation in their apartments. This is a difficult situation for medical personnel during the COVID-19 pandemic (Pradhan 2020). The stigma against medical personnel still occurs, but a movement to care for humanitarian fighters continues to emerge. Personal protective equipment (PPE) and aerosol boxes were donated by several communities to a number of hospitals in Indonesia. It is not only the task of the government but also the role of society to fight against COVID 19.

Stigma is a term that describes a situation or condition related to the point of view of something that is considered negative (Arboleda-Florez 2002). Stigma is understood as a social construction in which a distinguishing sign of social disgrace is attached to other people to identify and evaluate them. Stigma is common in the burden of disease (Wilsher 2011). Social stigma in the context of health is a negative relationship between a person or group of people who share certain characteristics and certain diseases (WHO 2020). Stigma can encourage people to hide illness to avoid discrimination, prevent people from seeking immediate health care, and prevent them from adopting healthy behaviors. The emergence of a negative stigma against medical personnel working in hospitals originated from the assumption that medical personnel who work in hospitals can transmit COVID-19 to people in the surrounding environment. Stigma hurts the person/group and has even more negative impacts on mental health compared to COVID-19 itself. The stigma that circulates in society can lead to discrimination, exclusion, and expulsion. Courtwright and Turner (in Craig et al. 2016) suggest that stigmatization is different from discrimination, as the former occurs with shame, while the latter involves exclusion. Stigmatization is seen as a complex process involving institutions, communities, and interpersonal attitudes. The purpose of this research is to find out the negative stigma attached to medical personnel, both from the family and society in the midst of the COVID-19 pandemic, to know the impact of negative stigma in the midst of the COVID-19 pandemic, and to know the solutions applied to prevent and overcome the negative stigma amid the COVID-19 pandemic.

2 METHODS

This research employed a qualitative research approach. Qualitative research is enacted based on the philosophy of interpretivism that is used to examine the condition of natural objects, where the researcher is a key instrument and the research results emphasize meaning rather than generalization (Sugiyono 2009). In qualitative research, what is used as the target of the study is the conditions of social life or society as a whole and comprehensive unit. This is related to the issues raised, namely regarding the socio-cultural problems of the community in the midst of the COVID-19 pandemic.

The data collection techniques used in this study were participatory observation, free, and in-depth interviews via video call and/or telephone and were supported by a literature review. The process of selecting informants was taken purposively, in which the selected informants were those who understood the research problem raised, so that the data was appropriate and did not deviate far from the research topic. This study involved 30 informants from 15 referral and non-referral hospitals for COVID-19 in the City of Surabaya. The data analysis in this study was carried out through three stages, namely making field notes, reducing data, and drawing conclusions. Through this abstraction, the prevailing social institutions in the area or community where the research is conducted were seen (Ahimsa-Putra 2009). Findings in the field were processed with data obtained from the literature and were presented in a descriptive ethnographic work (Denzin & Lincoln 2009).

3 RESULTS AND DISCUSSION

Pandemic checks and social stigma, as well as the strengthening and decentralization of COVID-19 medical facilities including testing, tracing, formal quarantining, and special care for COVID-19

Table 1. Informant's data

Profession	Number
Nutritionists	1
pharmacist assistant	1
Midwife	8
Medical Physics	1
Environmental Health	1
Nurse	17
Radiographer	1

patients must be ensured by a relatively large state budget and collaboration with the private sector for health management. The government of Bangladesh and some voluntary organizations should establish a Central Stigma Management Committee (CSMC) that would work with local committees that use relevant experts in the field and use official guidelines that can motivate people to support their families, communities/local residents, and their neighbors as a form of moral responsibility (Mahmud 2020).

Stigma is a strong social process characterized by labeling, stereotyping, and segregation that leads to loss of status and discrimination, all occurring in the context of power. Stigma also has an impact on the welfare of health workers because health workers live in a stigmatized condition (Nyblade 2019). The social stigma that appears in the community regarding medical personnel is the result of a lack of knowledge of COVID-19 (transmission of the disease, ways of handling, and prevention). The risk of transmission should be given by the local community or local officials to the community to avoid rumors or misinformation circulating so that the information circulating is correct information based on scientific evidence. Health workers at the forefront of handling COVID-19 cases have a high level of risk of being exposed to COVID-19. There are a lot of negative stigmas that Indonesian health workers receive from areas affected by the pandemic. Nurses often face a negative stigma since people think they may spread the virus. Health workers who are at the forefront of handling COVID-19 cases have a high level of risk of being exposed to COVID-19.

Apart from Indonesia, this also happens abroad; as newspaper reports in mid-March 2020, a female doctor working at Dhaka Medical College received an ultimatum from the owner of her neighboring apartment in Old Dhaka to leave her apartment or job, otherwise, they threatened to expel her from the building (Kamal 2020 in Mahmud 2020). The massive effects of the social stigma associated with COVID-19 are so concerning that previous social norms, social values, and relationships, as well as social capital, have been shattered in surprising ways.

This study involved medical personnel who worked in 15 hospitals in the city of Surabaya. The proportion of informants who participated in this study can be seen in the diagram below. The number of informants who participated in this study was 30 people. The 30 informants have jobs as nutritionists, assistant pharmacists, midwives, medical physics, environmental health, nurses, and radiographers. Medical personnel who live in Surabaya receive more of the negative stigma from their environment. This is due to the high number of COVID-19 patients in Surabaya, which is the highest in East Java.

Based on the results of interviews, some of the medical personnel experienced negative stigma. The negative stigma received by medical personnel comes from neighbors and even immediate family members. The attitudes shown are also of various kinds, such as discrimination, exclusion, and bullying, the worst of which is expulsion. The negative stigma in the form of bad treatment also came from neighbors, and this was experienced by DS (29). DS received unpleasant treatment because she was in charge of handling Covid-19 patients. This had an impact on her mother's shop, which no one visited. DS herself felt that it was okay to be excommunicated. DS's children are ostracized by their neighbors and friends, namely their neighbors and their friends' parents do not allow their children to play with DS's children. DS feels grateful because her son does not have an emotional attitude. That is, he accepts the treatment and prefers to stay at home and play with his younger siblings. DS and her family have never experienced evictions from their neighbors at home. DS's father was worried because he worked dealing with COVID-19 patients, but DS tried to explain, and finally, her father understood and accepted DS's explanation. DS explained to her father that she had to go home to take care of her family, and as long as she could maintain cleanliness and health, nothing would happen.

The neighbors around the house and those who live in the boarding houses experience a more unpleasant incident, as experienced by MM (30). MM was excommunicated by his boarding house neighbor for about one month. His neighbors were indifferent and did not communicate with MM, even the owner of the boarding house also isolated him. Before the pandemic, he chatted with his neighbors, while during the pandemic there was no opportunity to chat with his neighbors. MM performs a SWAB test at the hospital, which is renewed every three months. The SWAB test results showed that he was negative, indicating that he was not exposed to COVID-19. MM's boarding house neighbor who knew about it experienced a change in attitude from initially ignoring and isolating from MM to returning to normal.

The stigma that comes from the nuclear family is experienced by YC (32). The response of the family and the environment related to work as medical personnel working at the COVID-19 referral hospital is that food should not be shared, YC's cutlery or utensils are kept away from other family members to reduce direct contact. YC often drinks bottles of mineral water by not using the glass so that the glass he has used is not used by other family members.

YS (30) received a negative stigma from the community because her husband was suspected of being positive for COVID-19 and because, according to him, most people were ordinary people who did not understand health, so many viewed their family negatively. YS has also reduced direct contact with neighbors since the pandemic. The bad experience because of the negative stigma of society was obtained by YS when she wanted to buy antiseptic from her neighbor after the neighbor knew YS's name and address, the neighbor looked worried, then YS decided to buy somewhere else and didn't think too much about it. Some neighbors stay away from their families, but there are also those who still want to be in touch with them; for example, a neighbor wants to deliver an order that YS bought from the neighbor to his house. Negative stigma can even come from the workplace and from fellow health workers. The negative stigma of the work environment is experienced by DN (26). DN had the experience when he first guarded the COVID-19 room. None of his friends wanted to talk to him for fear of catching the virus, including the doctor who was on duty at the ER. DN experienced discrimination in the hospital after guarding the COVID-19 room. The impact of negative stigma from the environment and society causes anxiety and disrupts a medical professional's psychological condition. The impact of this negative stigma makes medical personnel reluctant to socialize with their neighbors and extended family, as was done by DN.

Facing this negative stigma, the hospital provides several policies, such as the rapid test, which is carried out regularly and even the SWAB test for medical personnel who need it, the use of PPE for medical personnel, as well as a place to stay in the hospital for medical personnel who want to stay in the hospital. Rapid tests are carried out periodically at the seven hospitals where the informant comes from. If the results are positive, then a SWAB test will be followed. There are seven hospitals that provide rooms, such as hotels or pavilions, that are used for medical personnel who do not want to return to their homes.

ST (27) often takes the rapid test in the hospital, and he is quite safe because he does not show symptoms such as cough, fever, and other symptoms. The symptoms experienced by COVID-19 patients are not very specific because many people are without symptoms. If the rapid test shows positive results, it is necessary to take a SWAB test. The workplace provides its own rapid test facilities to encourage testing and avoid transmission to worker's families. The most important test is rapid. If the results are negative, then the worker is safe and rest and vitamins are recommended. If the result shows a positive result, then a SWAB should be done. If the worker shows a positive rapid result, then they are not allowed to go home and must be isolated in the hospital.

4 CONCLUSION

Several medical personnel experienced a negative stigma during the COVID-19 pandemic. Medical personnel faced negative stigma from family, friends, neighbors, and even their immediate family members. The attitudes shown were also of various kinds, such as discrimination, exclusion, bullying, and expulsion from their residence. The impact of negative stigma from the environment and society causes concern and disrupts medical personnel's psychological condition. The impact of this negative stigma makes medical personnel reluctant to socialize with their neighbors and their extended families. Facing this negative stigma, the hospital provides several policies such as rapid and/or SWAB tests that are carried out regularly with medical personnel, especially for COVID-19 teams.

REFERENCES

Ahimsa-Putra, H.S. 2009. "Paradigma Ilmu Sosial-Budaya : Sebuah Pandangan." Seminar paper, Universitas Pendidikan Indonesia, Bandung, 7 Desember 2009.

Arboleda-Florez, J., 2002. "What causes stigma?", *World Psychiatry*, 1(1), hlm.25–26. https://www.ncbi.nlm.nih.gov/pmc/articles/PMC1489829/.

Craig G. M, A. Daftary, N. Engel, S. O'Driscoll, and A. Ionnaki. 2016. *Tuberculosis Stigma as a Social Determinant of Health: A Systematic Mapping Review of Research in Low Incidence Countries*. Denmark: Science Direct.

Denzin, N.K. dan Lincoln, Y.S. 2009. *Manajemen Data dan Metode Analisis. Handbook Of Qualitative Research*. Yogyakarta: Pustaka Pelajar.

Mahmud, Ashek and M. Rezaul Islam. 2020. *Social Stigma as a Barrier to COVID-19 Responses to Community Well-Being in Bangladesh*. Switzerland: Springer Link.

Nyblade Laura, Melissa A. Stockton, Kayla Giger, Virginia Bond, Maria L.Ekstrand, Roger Mc Lean, Ellen M. H. Mitchell, La Ron E. Nelson, Jaime C. Sapag, Taweesap Siraprapasiri, Janet Turan, and Edwin Wouters. 2019. *Stigma in Health Facilities:Why It Matters and How We Can Change It.* Washington DC: BM Medicine.

Pradhan Ravi Ranjan and Richa Nepal. 2020. *Stigmatization Towards Healthcare Worker During the COVID-19 Pandemic in Nepal*. Nepal: Nepalese Medical Journal.

Sugiyono. 2009. *Metode Penelitian Kuantitatif, Kualitatif, dan R&D*. Bandung: Alfabeta.

Varshney Mohit, Jithin Thomas Parel, Neeraj Raizada, and Shiv Kumar Sarin. 2020. *Initial Psychological impact of COVID-19 and its correlates in Indian Community: An Online (FEEL COVID) Survey*. New Delhi: Plos One.

Wilsher, E. J. 2011. *The impact of Neglected Tropical Diseases, and their associated stigma, on people's basic capabilities*. Durham University.

Community Empowerment through Research, Innovation and Open Access – Sayono et al (Eds)
© 2021 Copyright the Author(s), ISBN 978-1-032-03819-3

Character values in "curriculum 2013" in a typical traditional house in Batang-Batang Sumenep village as a response to socio-cultural changes

M.R. Gunawan*
Universitas Negeri Malang, Malang, Indonesia

ABSTRACT: Currently, typical traditional houses in Batang-Batang Sumenep have undergone changes along with the development of technology both in terms of decorative design, layout, and home models. Therefore, problems now exist because certain traditional philosophical values are now being excluded. The method used in this research was a descriptive qualitative method. The results showed that there are five values of character education in traditional houses that can be used as alternatives to address social and cultural changes. Based on the findings of the five samples of Batang-Batang village houses, the shape of the Bangsal house and the findings of the related five main character values were observed, namely: religiousness, nationalism, independence, mutual cooperation, and integrity. These values were integrated into the curriculum. Character values in traditional houses are very useful for students as they strengthen their nation's character.

Keywords: traditional houses, change, character values

1 INTRODUCTION

"Curriculum 2013" is the latest curriculum that began to be applied to the new school year in 2013. This curriculum has an important role in imposing the strengthening of character education in elementary schools, middle schools, high schools and even in higher education. According to Hartono (2017) the main characteristic of curriculum 2013 are integrating attitudes and skills into one lesson. Therefore, character values are integrated into every process of learning the curriculum. To build character, this curriculum continues to develop the good characters students already have (Sayono et al. 2015). In response to this, the president of the Republic of Indonesia issued a presidential regulation (Perpres, 87 2017) for strengthening character education in article 1, aiming to build and equip a golden generation of Indonesia in 2045 with the soul of Pancasila and the education of good character values to face change in the future.

One of the options for this in Madura island is historical heritage in the physical form of traditional houses. However, the philosophical values of traditional houses in Madura have been neglected, one of which is the character value. Character is very necessary to build the nation's generational identity, considering that the current era of young people in Indonesia, especially in Madura, is experiencing a degradation of character and morals. According to data (profiles of Indonesian children, 2019), it is known that in 2017 the number of child perpetrators of criminal acts reached 3,479, and in 2018 it reached 3,048. In addition, in 2018, in Madura, there were cases of violence against high school teachers. This can be seen in the news (CNN Indonesia 2018). There is a case of beating an art teacher by suspected students in Sampang, Madura, to death. This is a severe problem.

The researcher offers an alternative solution, namely the investigation of the character depicted in the typical traditional houses in villages in Batang-Batang Sumenep Madura. The value of

---

*Correspondin author: riskigunawan806@gmail.com

DOI 10.1201/9781003189206-7

character education is a valuable area for study because in the era of technology and information, technology can impact the influx of outside cultural influences. According to Kusmayadi (2017), external cultural values lead to dehumanization, namely that humans are no longer valued because they are more concerned with modern technological machines. In addition, cases of degradation of character and morals occurring among adolescents are increasingly prevalent. The purpose of this research was to explore the character of Curriculum 2013 vis-à-vis traditional houses in a typical village of Sumenep as a response to socio-cultural changes. This research is comprised of three sections: (1) an overview of traditional houses of the Batang-Batang village, (2) Curriculum 2013, which grounds character, and (3) the contents of character.

2 METHOD

This research employed qualitative and historical methods. According Prastowo (2012), the approach is treating the object to be studied. This research was classified as descriptive qualitative research in the form of written words. Data was collected from five traditional houses in Batang-Batang village. Traditional houses were chosen based on the criteria of houses that have not been renovated. After obtaining the data, the researchers conducted field observations to each house and conduct interviews with the community. The data obtained was analyzed using a historical approach by verification (of data source criticism, source validity) and interpretation to be used as historiography.

3 RESULTS AND DISCUSSION

3.1 *Description of a typical traditional house in Batang-Batang Village*

The research is located in Batang-Batang Village, Batang-Batang Sumenep District, Madura. According to the Sumenep Center for Statistics (2020), the boundary of Batang-Batang District is in the north bordering the Java Sea, south of Gapura, east of Dungkek, and west of Batu Putih. Batang-Batang village itself is divided into two, namely Batang-Batang Laok (south) and Batang-Batang Daya (north). The location for data collection is in Batang-Batang Daya village (north) to be precise was in the Sedung hamlet and Jeruk Porot hamlet. Traditional houses in each area have their own uniqueness and characteristics. Traditional houses according Sumardiyanto (2019) are cultural products of the community. The social values that have been maintained and carried out by the community will be reflected in elements of traditional buildings. According to Susilo (2018), the definition of a traditional house tends to be interpreted as a "place to live" (*Panggenan* or *Panggonan*, place). The traditional houses in Batang-Batang Village are included in the type *Bangsal* houses that have long been occupied by residents, the characteristic of this house can be seen from the shape of the roof which tends to have horns.

Traditional house architecture is much influenced by the religion, customs, and beliefs of local communities which are also influenced by climat conditions and location (Rumiawati & Prasetyo 2013). Traditional architecture usually refers to the form of traditional houses and residential buildings made in a community environment with certain customs (Rosyadi 2015). The climate in Madura is similar to other areas in Indonesia, which have tropical climates, because of its location in the lowlands and the dry land. These conditions also affect the style of the buildings in which people in Madura live. There are at least four forms of traditional houses in Madura seen from the shape of the roof, namely *trompesan, pegun, pacenan*, and *Bangsal*.

Traditional house in the Batang-Batang Sumenep village have the characteristics of the type of roof of the house *Bangsal*. According to Asmarani et al. (2016), the roof of the house is the development of the roof of *joglo* in Javanese architecture. The type of roof *joglo* adapted is *joglo Lawakan* and *joglo Sinom*. The difference that the sides and back are clipped. In Batang-Batang, everyone has the same type of house, namely the roof of *Bangsal*. It should be noted that the characteristics of the *Bangsal* house have a horn on the top, but those found in the Batang-Batang village in the flower area do not have horns. This means the construction of a traditional house is not too influenced by Chinese culture, considering that the traditional houses around the Sumenep palace had horns on the top.

The floor of *Bangsal* has an average height of 40 cm with forming *de'ondek* (terraces) with the lapped bottom having a height of 20 cm and top 20 cm, while its basic form consists of square and rectangular. In general, the *Bangsal* houses in Batang-Batang Sumenep village only have two rooms, namely the inner and outer spaces. The outer space, known as *amper,* is a place to receive guests while the inner space is private, for sleeping, and storing agricultural products and other valuables. The floors in the *Bangsal* village of Batang-Batang house are generally original material made of clay. Subsequent developments of some parts were renovated made of plaster in the form of black cement. In general, in Madura, the use of building materials such as wood and bamboo are very dominant, and most of the wood gets special treatment such as teak wood and jackfruit wood, because of very close relationships to values, norms, and customs. The walls in the traditional houses of Batang-Batang Sumenep village use more material *beto pote* (white stone). White stone materials are used because *beto pote* are the source of the material which is abundant and until now its resources are still traded. The window type in this *Bangsal* house is a window with two leaves, and the door has *Kupu Tarung* door type.

3.2 *The character of curriculum 2013*

This curriculum is the latest curriculum to take effect in the new school year 2013. The development of the 2013 curriculum is expected to produce a productive, creative, innovative generation. According to Permendikbud No. 35 of the year 2018 on changes to Permendikbud No. 58 of the year 2014 on curriculum 2013 in Junior High School/Madrasah Tsanawiyah: "Curriculum 2013 aims to prepare Indonesians to have the ability to live as a person and citizen who believe, productive, creative, innovative, and affective and able to contribute to the lives of people, countries, countries, and civilizations of the world".

As a result, the 2013 curriculum applies character content at all levels of education. Character is very useful for the nation's next generation in the future as a booster of the nation's identity at a time when social and cultural change is rapidly moving. According to the "Guidelines for Strengthening Character Education 2017," as the foundation of curriculum 2013, the national movement of character education has begun in 2010 which has spawned a pioneer school capable of carrying out character formation contextually in accordance with the potential of the local environment. The government realizes that the National Movement for Mental Revolution that strengthens character education should be implemented by all schools in Indonesia, not just limited to target schools. In response, in 2017, the president of Republic Indonesia issued presidential regulation 87 of the year 2017 concerning strengthening character education that applies to all levels of education in Indonesia.

3.3 *Tradisonal house character value content*

The character in today's era is very much needed to form a person with a noble character. Moral degradation in Indonesian adolescents is a very serious problem because, if left uncorrected, it would be like a time bomb. Therefore, as an educator who is aware of the problem, acts immediately as the savior hero of the generation of the nation, one of which provides character-based learning. Character education is a necessity in an effort to face the various challenges of character shifting faced today (Komara 2018). Character education refers to the values of the character of the nation, contained in Permendikbud No. 20 of 2018 Article (2):

a. PPK is implemented by applying Pancasila values in character education mainly covering religious values, honest, tolerant, discipline, hardworking, creative, independent, democratic, curiosity, national spirit, love of the homeland, appreciating achievements, communicative, peace of mind, fondness for reading, environmental care, social care, and responsibility.

b. The value as referred to in paragraph (1) is the embodiment of 5 (five) interconnected main values namely religiosity, nationalism, self-reliance, mutual cooperation, and integrity integrated in the curriculum.

To find out what values are contained in the *Bangsal* house in Batang-Batang Sumenep Village, researchers refer to the values of national character, found in Permendikbud Number 20 of 2018 Article (2) which has been summarized into only five values. The following is a description of the five-character values found in the traditional houses of Batang-Batang Sumenep Village.

3.3.1 *Religiousness*
There is no doubt about the religious value of the Madurese people. They believe in the power of God so that they can express it in their daily lives and one of these expressions is the traditional Madurese house design. Before making a house, the Madurese community has the habit of counting the days of goodness so that the house built becomes a blessing for the home owner, right before the construction process, they dig the foundation for planting four pedestals to support the house poles. Before installing the *umpak* (pedestals), a salvation is held as a form of gratitude and to ask for prayers so that those who occupy the prospective house will receive safety. According to Dalyono & Lestariningsih (2017), religiousness is the attitude and behavior of being obedient in carrying out the teachings of the religion they adhere to, tolerant of the implementation of other religions, and living in harmony with followers of other religions.

3.3.2 *Nationalism*
As another area in Indoensia will surely tip the value of nationalism, Madura people have long been promoting the value of nationalism as a manifestation of their love of their Indonesian homeland. In 1900, Indonesia was not independent but the rehabs used to respect *Buppa'* (Father), *Bhabhu'* (Mother) *Ghuru* (Teacher), *Rato* (Leader) turned out to have a profound impact on subsequent developments. Respecting *Rato* means respecting the leaders/bureaucracy of the authorities, Until Indonesia is independent, whoever leads the Madurese community will always be respected. This philosophy is manifested in the physical buildings of Madurese houses, the traditional buildings on average will never face the main road because they respect the nobles who will pass on the highway.

3.3.3 *Independence*
The Madurese are very independent, resilient, and hardworking. They have the desire to migrate to get a job and guarantee a happy life. It seems that the independence of the Madurese community had long been formed when they were teenagers, and this cannot be separated from where they live. The Madurese community lives in groups with their own family members known as *taniang lanjeng*. *Tanian lanjeng* contains a group of families. The residence will face south and the kitchen and drums are in the south. The *pakeban* (bathroom) is outside the house, while *kobhung* (place of worship) is located in the west. The philosophy in the house with the *tanian lanjeng* system is increasingly western looking and sacred. When boys turn into teenagers, they are allowed to sleep outside in places of worship. This is because they are used to keeping families out of danger from the outside and are taught to take responsibility.

3.3.4 *Cooperation*
The spirit of *gotong royong* (cooperation) has been ingrained in the Madura people. They help each other work together. The process of making a house is done by means of deliberation, collecting several families to take advice regarding the person's intention to build a new house. This is in accordance with the opinion of Dewantara (2017) on mutual cooperation being a dynamic principle, even more dynamic than kinship. Cooperation describes a joint effort and mutual assistance toward a common interest.

3.3.5 *Integrity*

The unity or integrity is seen in the system of conscience of the family who lives in one *tanian lanjeng*. They help each other and try to get along despite having to live in a yard with their own family. The community is also friendly and hospitable to the presence of guests from other races or cultures, and guests should be welcomed like kings. Homes are sometimes constructed for guests who have been forced to stay a long time.

4 CONCLUSION

This research has shown that the 2013 curriculum is based on character education and is presented as a character booster to the nation's growing generation. In response to the policy, researchers suggest the government consider the character value of the typical traditional house of a Batang-Batang Sumenep Madura village. The character values obtained are the manifestation of five main values that are interrelated, namely: religiousness, nationalism, independence, mutual cooperation and integrity, which are integrated into the curriculum.

REFERENCES

Asmarani, I. K. & Antariksa, & Ridjal, M. A. 2016. Tipologi Elemen Arsitektur Rumah *Bangsal* di Desa Larangan Luar Pamekasan Madura. *Jurnal Tesa Arsitektur*. 14 (1), 10–22.
Badan Pusat Statistik Kabupaten Sumenep. 2020. *Kecamatan Batang-Batang Dalam Angka 2020.*
Dalyono, B., & Lestariningsih, D. E. 2017. Implementasi Penguatan Pendidikan Karakter di Sekolah. *Jurnal Bangun Rekaprima.* 3 (2), 33–42.
Dewantara, W. A. 2017. *Alangkah Hebatnya Negara Gotong Royong (Indonesia dalam Kacamata Soekarno).* Yogyakarta: PT Kansius.
Dharmawan, N.S. 2014. *Implementasi Pendidikan Karakter Bangsa pada Mahasiswa di Perguruan Tinggi.* Paper from Pembinaan Pendidikan Karakter bagi Mahasiswa PTS, Lingkungan Kopertis Wilayah VIII Tahun 2014.
Hartono, Y. 2017. Model Pembelajaran Nilai-Nilai Karakter Bangsa di Indonesia dari Masa ke Masa. *Jurnal Agastya.* 7 (1), 34–48.
Kerjasama Kementrian Pemberdayaan Perempuan dan Perlindungan Anak dengan Badan Pusat Statistika. 2019. *Profil Anak Indonesia.* Jakarta: KPPPA.
Komara, E. 2018. Penguatan Pendidikan Karakter dan Pembelajaran Abad 21. *Jurnal Sipatahoenan.* 4 (1), 17–26.
Kusmayadi, Y. 2017. Hubungan Antara Pemahaman Sejarah Nasional Indonesia dan Wawasan Kebangsaan dengan Karakter Mahasiswa (Studi pada Mahasiswa Pendidikan Sejarah FKIP Universitas Galuh Ciamis). *Jurnal Agastya.* 7 (2), 1–19.
Pedoman PPK. 2017. *Konsep dan Pedoman Penguatan Pendidikan Karakter.* Pusat Analisis dan Sinkronisasi Kebijakan Sekretariat Jendral Kementrian Pendidikan dan Kebudayaan.
Peraturan Presiden (perpres) 87 tahun 2017 mengenai penguatan Pendidikan karakter
Permendikbud Nomor 35 Tahun 2018 Tentang Perubahan atas Permendikbud Nomor 58 Tahun 2014 Tentang Kurikulum 2013.
Prastowo, A. 2012. *Metode Penelitian Kualitatif dalam Perspektif Rancangan Penelitian.* Yogyakarta: Ar-Ruzz Media.
Rosyadi, 2015. Trdisi Membangun Rumah dalam Kajian Kearifan Lokal (Studi Kasus pada Masyarakat Adat Kampung Dukuh). *Jurnal Patanjala.* 7 (3), 415–430.
Rumiawati, A., Prasetyo, YH. 2013. Identifikasi Tipologi Arsitektur Rumah Tradisional Melayu di Kabupaten Langkat dan Perubahannya. *Jurnal Permukiman* VIII (2): 78–88.
Sayono, J., & Nafi'ah, U., & Wijaya, N. D. 2015. Nilai-Nilai Pendidikan Karakter dalam Dongeng Gagak Rimang. *Jurnal Sejarah dan Budaya.* 9 (2), 236–256.
Sumardiyanto, B. 2019. Pengaruh Renovasi Terhadap Makna Rumah Tradisional Masyarakat Jawa, Kasus Studi: Kotagede Yogyakarta. *Jurnal ARTEKS.* 3 (2), 99–114.
Susilo, A. G. 2018. Model Tata Masa Bangunan Rumah Tradisional Ponorogo. *Jurnal Lingkungan Binaan Indonesia.* 7 (1), 60–67.

Community Empowerment through Research, Innovation and Open Access – Sayono et al (Eds)
© 2021 Copyright the Author(s), ISBN 978-1-032-03819-3

Fun and creative learning using discovery learning in Islamic economics: A Technological approach

V.A. Qurrata*, S. Merlinda & R.D.L. Puteri
Universitas Negeri Malang, Malang, Indonesia

ABSTRACT: The purpose of the present study was to investigate the implementation of discovery learning on Islamic economics. This study used classroom action research. The subject was 25 undergraduate students who were enrolled Islamic economics majors, and data was collected using a questionnaire, observation, tasks, and documentation. The data was analyzed descriptively. The results showed that the participants never used discovery learning for the class activity before. This new experience had a positive impact on the participants' perception on Islamic economics learning. Discovery learning was found to be more practical compared to the traditional method. Moreover, this study found that the participants were attracted to learning the idea in discovery learning through direct interaction with society.

Keywords: discovery learning, Islamic economics, classroom action research

1 INTRODUCTION

Learning models are varied and have characteristics. One of these models is discovery learning, which is a model used to solve problems by students independently under teacher supervision (Indriyanti & Prasetyo 2018). Furthermore, discovery learning is a cognitive learning method that requires teachers to be more creative in creating situations that can make students learn to discover their knowledge actively. Through this method, students will try to learn and find conclusions from a material. Through this method, it is hoped that students will not only listen passively to the lecturers but will be more active in seeking new knowledge.

The learning process through a discovery learning model using a scientific approach can improve students' critical thinking skills and learning outcomes (Balım 2009). Learning outcomes using the discovery learning method could increase students' average scores in mathematics learning outcomes compared to conventional methods (Sari & Sukartiningsih 2014). The results of implementing the discovery learning model were also stated by Joyce and Weil (2009), using this method, students can increase activeness and participation in learning activities. This condition is reflected in the courage of students to express, argue, accept other people's opinions and work together in groups. The dominance of specific student groups is no longer visible, and various activities are carried out by each student with full responsibility such as in the presentation of the material. Additionally, learning through this method is also supported by the improvement of every aspect of assessment: both in aspects of the process and in learning outcomes (with the achievement of minimum completeness criteria).

Based on the description above, the learning model that is considered capable of empowering social attitudes, cognitive knowledge and science process skills is the discovery learning model. This learning model should be considered and implemented in learning Islamic Economics. Islam distinguishes between economics and the economic system. In the general definition, the system is a complex whole, namely an arrangement of things or parts that are interconnected, whereas

---

*Corresponding author: vika.annisa.fe@um.ac.id

 DOI 10.1201/9781003189206-8

science is the knowledge that is formulated systematically. The system can be defined as any rule born from a worldview or a specific creed that functions to solve and overcome the problems of human life, one that explains how to solve, maintains and develop these solutions. At the theoretical and conceptual level, teaching Islamic Economics is limited to curriculum sources and materials. The books used in learning activities do not effectively encourage students to find concepts and express their ideas (Amalia & Al Arif 2013). The Islamic Economics material is challenging to understand and does not encourage students to carry out science-based learning activities (Febrianto & Wahyuningsari 2017).

Islamic economics is a new science that is still common to students. Islamic economics has many foreign terms in Arabic, which are difficult for students to memorize. Furthermore, the concept in Islamic Economics is still new, especially for students who are not too steeped in religious knowledge. Therefore, the discovery learning method is important to be used in learning Islamic Economics in a fun way and makes it easier for students to understand this science. This research wants to find out how effective this method can be in teaching students to understand Islamic Economics.

2 METHODS

The design of the implementation of learning innovation in the discovery learning model was carried out on subjects whose substance contains case study elements that require a solution based on existing theory and require a legal basis. Wenning (2011) stated that the steps of discovery learning are observation, manipulation, generalization, verification, and application. Students face phenomena that interest them in the module's observation stage and respond to them. This is in agreement with Lavine (2005), which stated that clinical cases presented in guided discovery learning function to focus on real problems and increase relevance and motivation to master science basics.

This module's outstanding characteristic is that in every activity, it places great emphasis on group cooperation in discovering concepts, not on individuals. Modules are the best way for students to actively study together with friends. At the same time, the lecturer checks intensively and aids students who have difficulty studying modules individually (Swaak et al. 2004). The following is a picture of the discovery learning development design implemented in the Islamic Economics course.

Based on the image above, there are 4 phases in discovery learning specifications. First, the material is discussed and the groups divide. Second, discussions are conducted by students in one group. This discussion aims to finalize the concept. Then, students also conduct interviews with the community and experts according to each group theme. The results of the interview are then used in making the video (vlog). Third, the video packaged in the form of a vlog is uploaded to the IG Department of Development Economics and UM E-learning. Next, the video is presented and discussed with other groups. Then, the teacher gives a case to be solved by the group concerned. Fourth, students compile papers based on the material that has been discussed and the explanations and conclusions of the speakers. The teacher provides a review and additions (if needed) to the modules prepared by students from the papers prepared. Next, the lecturer arranges modules on each related theme, then uploads them to the e-learning system. Fifth, in addition to the value, the lecturer will task the student with whether they can develop concepts and calculations according to the software theme. Sixth, validation, which is carried out by material experts in each field.

After the e-module output is compiled, there is the need for further validation. Validation is carried out to test the discovery learning based Islamic economics module's feasibility, so experts are needed to do this. Validation is a process or validation of the module's conformity to the needs (Kamel 2014). Furthermore, Daryanto (2013) states that validation can be done by asking for help from experts who master the studied competencies. Thus, the requirements for a validator are that they must hold bachelors or masters degree titles in their field as needed. There are two validators in the assessment of this module, namely material experts in: (i) financing, (ii) zakat, (iii) waqf,

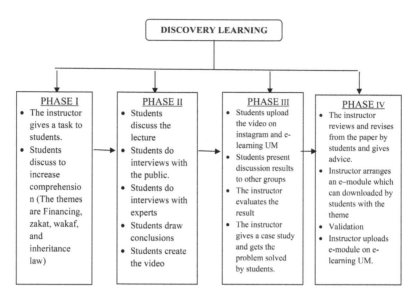

Figure 1. The phase of discovery learning's development plan.

and (iv) inheritance. Islamic Economics lecturers carry out the material validator. The purpose of material validation is to determine the feasibility and adequacy of the material presented. In preparing the material, it must pay attention to the depth and breadth of material coverage. The material's breadth describes how much material is included, while the depth of the material concerns the details of the concepts contained in it that students must study.

The data analysis was done by describing the research results. Based on the observation instruments, it can be seen that the teacher's and students' activities were shown using frequency, percentage, and mean. Moreover, based on the observation results during the learning implementation, the teacher's activities related to giving the stimulus, leading the problem identification, and collecting the data combined with aspects of the scientific approach, namely observing, asking questions, exploring, associating, and presenting. Meanwhile, the students' learning results were given scores to get the learning competence made into grades.

3 RESULTS AND DISCUSSION

There are several points to assess creative thinking in this research, such as generating many ideas in various categories, having new ideas to solve the problem, and having the ability to solve the problem in detail (Kim 2006). Thus, it has shown in Table 1 below.

Table 1. Descriptive analysis result.

Component	Pre-Test	Post-Test
Maximum Score	70	90
Minimum Score	20	50
Average	50.77	70.25
N	25	25

In this study, the students must learn with problem-solving activities, investigation, discovery, small-group work, learner-centered education, and using lecturer guidelines. True learning is learning that examines human problems in relation to the world (Bruner 1961). Discovery learning needs students to solve their own problems with the lecturer as a "trainer" for students. Students

Table 2. Students creative thinking ability test.

Component	Pre-test	Post-test
Less creative	15	0
Quite creative	5	10
Creative	5	10
Highly Creative	0	5
N	25	25

are asked to "think" more and create rather than "quote" material (Schoenfeld 1980). Investigation is also important in this process because the lecturer can submit the lesson as material for students' thinking which encourages them to re-examine their previous thoughts. Students are free to express their thoughts (Simamora et al. 2019). Moreover, investigating is the process of discovering new things. In the discovery process, students must be involved in the discovery of various concepts and principles through problem-solving or the results of abstracting various cultural objects (Schoenfeld 1987).

However, this method is more effective using small groups compared to large groups. Schoenfeld (2013) said that ideas formed by individuals were often built and refined in collaboration with others. This implies that all cognitive, even high-level work in humans, starts from culture, and that means students should learn through interaction with adults and more capable peers. Teachers need to implement learning strategies that allow students to interact with their friends. In discovery learning, students are encouraged to learn on their own independently. Students are actively involved in discovering various concepts and principles through problem-solving or the results of the abstraction of various cultural objects. The teacher encourages and motivates students to gain experience by doing activities that allow students to discover mathematical concepts and principles for themselves. This learning arouses curiosity and fosters motivation in students to work until they find the answer. Students learn to solve problems independently with thinking skills because they must analyze and manipulate information (Simamora et al. 2019).

In the guided discovery learning model, the teacher acts as a guide. This guidance is needed to anticipate negative things such as cognitive overload, potential misunderstandings, and the teacher's difficulties in detecting problems and misunderstandings. Alfieri et al. (2011) stated that discovery learning without assistance has not benefited students, while feedback, examples of successful work, scaffolding, and teacher explanations such as reinforcement will benefit student learning achievement. Opportunities for constructive learning may not arise when students are left without help. The guided activity of the teacher has a scaffolding to help students. The activities given by the teacher require students to explain their ideas and ensure that these ideas are accurate by providing timely feedback from the teacher. The activities provided by the teacher provide examples of work on successful tasks. Table 2 presents the results of a student creative thinking ability test.

Participants' perceptions regarding the discovery learning method are more effective than ordinary learning methods or lectures. More than 90% of the participants stated that the discovery learning method is more effective than the usual method. Only 8 percent of participants disagreed. This is in line with Tran's (2014) research, which concluded that the discovery learning model is more effective than the traditional learning model. The difficulties in the discovery learning method are divided into five parts, namely: (i) understanding the material from related references; (ii) the interview process with experts; (iii) the interview process with the community; (iv) the process of presenting the results in the form of making videos and vlogs; (v) and the delivery of results. The most dominant difficulty faced by participants is presenting the results in the form of making videos and vlogs.

Making videos and vlogs is the hardest part. However, nearly one-third percent of participants stated that the presentation process in the form of video and vlog was the most interesting thing in the discovery learning process. Discovery learning is one method that is attractive and attracts participants. This is evidenced by 80 percent of the participants expressed their agreement/interest that if this method was applied, there were other courses.

4 CONCLUSIONS

Discovery learning can encourage students' creative thinking abilites in learning and teaching strategy subjects. There are huge differences in the pre- and post-test scores after discovery learning is implemented. This research implies that using discovery learning models is one solution to enhance students' creative thinking abilities. Moreover, this learning model is fit to be implemented in various fields of education. Meanwhile, if this model is applied to several exact courses, it should be combined with other models to increase students' understanding. This combination is done because students have different abilities and ways of absorbing knowledge. The combination is done, for example, by conducting intermittent lectures, testing individual understanding through quizzes to determine the effectiveness of learning outcomes and as learning evaluation materials.

REFERENCES

Alfieri, L., Brooks, P. J., Aldrich, N. J., & Tenenbaum, H. R. 2011. Does discovery-based instruction enhance learning? *Journal of Educational Psychology*, 103(1), 1–18. https://doi.org/10.1037/a0021017

Amalia, Euis and M. Nur Rianto Al Arif. 2013. Kesesuaian Pembelajaran Ekonomi Islam di Perguruan Tinggi Dengan Kebutuhan SDM Pada Industri Keuangan Syariah di Indonesia. *INFERENSI: Jurnal Penelitian Sosial Keagamaan* Vol. 7 No. 1

Balım, A. G. 2009. The effects of discovery learning on students' success and inquiry learning skills. *Eurasian Journal of Educational Research*, 35(35), 1–20.

Bruner, J. S. 1961. The Act of Discovery. *Harvard Educational Review*, 3(1), 21–32.

Daryanto. 2013. *Menyusun Modul: Bahan Ajar Untuk Persiapan Guru Dalam Mengajar*. Yogyakarta: Gava Media.

Febrianto, Nur Fitroh and Wahyuningsari. 2017. CLC (Centered Learning Circular): Metode Revitalisasi Keilmuan Ekonomi Islam Untuk Kalangan Mahasiswa. *al-Uqud: Journal of Islamic Economics*, Volume 1 Nomor 1, Januari 2017. Hal 83–94

Indriyanti, Rita and Zuhdan Kun Prasetyo. 2018. Improving the Experiment Report Writing Skills of Fifth Graders Through the Discovery Learning Method. *Jurnal Prima Edukasia*, 6 (1), 2018, 104-102-110

Joyce, Bruce & Weil. (2009). *Models of teaching*. New Jersey USA: Pearson Education, Inc, Publishing as Allyn& Bacon.

Kamel, Abdelrahman. 2014. The Effect of Using Discovery Learning Strategy in Teaching Grammatical Rules to first year General Secondary Student on Developing Their Achievement and Metacognitive Skills. *International Journal of Innovation and Scientific Research* Vol. 5 No. 2 Jul. 2014, pp. 146–153.

Kim, K. H. 2006. Can We Trust Creativity Tests? A Review of the Torrance Tests of Creative Thinking (TTCT). *Creativity Research Journal,* 18(1), 3–14. https://doi.org/10.1207/s15326934crj1801_2

Lavine, Robert. 2005. Guided Discovery learning with Videotaped Case Presentation in Neur biology. *JIAMSE*. Vol. 16, 4–7.

Sari, V. N., & Sukartiningsih, W. 2014. Penerapan model discovery learning sebagai upaya meningkatkan kemampuan menulis teks cerita petualangan siswa kelas IV sekolah dasar. *Jurnal Penelitian Pendidikan Guru Sekolah Dasar*, 2(2), 1–10.

Schoenfeld. A. H. 1980. Teaching Problem-Solving Skills. *The American Mathematical Monthly*, 87(10), 794–805. https://doi.org/10.2307/2320787.

Schoenfeld, A. H. 1987. Polya, Problem Solving, and Education. *Mathematics Magazine*, 60(5), 283–291.

Schoenfeld, A. H. 2013. Reflections on Problem Solving Theory and Practice. *The Mathematics Enthusiast*, 10(1,2), 9–32.

Swaak, J., Jong, T,D. & Joolingen, W.R. 2004. The effect of discovery learning and expository instruction on the acquisition of definitional and intuitive knowledge. *Journal of Computer Assisted Learning*. 20: 225234

Tran, Trung. 2014. Discovery learning with the Help of the Geogebra Dynamic Geometry Software. *International Journal of Learning, Teaching and Educational Research (IJLTER)*. Vol. 7 (1), pp. 44–57.

Wenning. 2011. Levels of Inquiry Model of science teaching: Learning sequences to lesson plans. *Journal of Phisics Theacher Education Online*, 6(2), 17–20.

Community Empowerment through Research, Innovation and Open Access – Sayono et al (Eds)
© 2021 Copyright the Author(s), ISBN 978-1-032-03819-3

Do they not seem normal? Indonesian teachers reading gender biased folktales

E. Eliyanah* & A. Zahro
Universitas Negeri Malang, Malang, Indonesia

ABSTRACT: Gender responsive education needs gender sensitive teachers as they will be able to provide teaching and learning activities that critically engage with the social construction of genders and address gender biases. Our research examines teachers' sensitivity towards gender biases in teaching materials, in this case, in the written version of Indonesian folktales. We asked 24 teachers from Medan, Malang, and Ambon to fill out a reception questionnaire on the characters and character development. Our analysis on the teachers' responses shows that these teachers tend to overlook gender biases in the sampled folktale. They barely noticed the biased characterization and treatment of the male and female protagonists. In doing so, teachers risk sustaining the normalized gender biases and maintaining practices. This finding implies the pressing need for awareness and gender training among teachers in Indonesia to ensure the implementation of gender responsive education.

Keywords: folktale, gender bias, reading, gender responsive, education

1   INTRODUCTION

Gender responsive education needs gender sensitive teachers. This is because teachers have great potential in inspiring and empowering their students (Frei & Leowinata 2014; UNESCO 2017). Thus, in order to instill values of gender equality in students, the need for gender sensitive teachers is unquestionable. Gender sensitive teachers will be able to provide teaching and learning activities which critically engage with the social construction of genders and address gender biases. The opposite generally, whether intentionally or unintentionally, will likely maintain normalized gender biases. Thus, teachers should have the capacity to integrate gender perspectives into their teaching (The Global Partnership for Education and UNGEI 2016).

Our research examines teachers' sensitivity towards gender biases in teaching materials, in this case, in the written version of Indonesian folktales. We involved 24 teachers from Medan, Malang and Ambon in filling out a reception questionnaire. Our analysis of the teachers' responses shows that these teachers tend to overlook gender biases in the sampled folklore. Regardless of the teachers' gender, age, and cultural origin, they barely noticed the biased characterization and treatment of the male and female protagonists. In doing so, teachers risk sustaining the normalized gender biases, which tend to be disadvantageous to women, and maintaining practices which tend to discriminate against women. This finding implies the pressing need for awareness and gender training among teachers in Indonesia to ensure the implementation of gender responsive education.

---

*Corresponding author: evi.eliyanah.fs@um.ac.id

Table 1. Research participants.

Region	Gender	Age	Total
Medan	Female	<35	5
		<45	2
		>45	1
	Male	<35	2
		<45	0
		>45	0
Malang	Female	<35	3
		<45	0
		>45	3
	Male	<35	2
		<45	2
		>45	0
Ambon	Female	<35	0
		<45	0
		>45	2
	Male	<35	0
		<45	1
		>45	1
Total			24

2 METHODS

This research is part of a larger research project on the representation of gender in the written version of Indonesian folktales. Our findings on the representations of female heirs in the contemporary written version of Indonesian folktales show that despite the dynamic nature of folktales, the characterizations of female heirs, or princesses, are still anchored in their looks and in a cult of domesticity (Zahro et al. 2020). In this paper, we would like to report our findings on how elementary school teachers in Indonesia, who are part of the target users of the written folklore, receive the stories. Data was collected through distribution of a questionnaire. The collected data was then analyzed using discourse analysis.

The stories that we studied are intended for young readers. Such intentions are written in the cover page of each story book. Thus, we assume that elementary school teachers have important roles in promoting the books to the intended readers, including using the books as part of their teaching materials. The books are electronically published and available for free download from the website of *Badan Pengembangan dan Pembinaan Bahasa* (Institute for Language Development and Preservation), from the Ministry of Education and Culture. For this reception study, we randomly selected *Kisah Dewi* Samboja: *Cerita Rakyat dari Jawa Barat* (The Story of Princess Samboja: A Folktale from West Java), written by Nia Kurnia (2016), to be read by our participants. The story revolves around the trials and tribulation of *Dewi* Samboja (Princess Samboja), a female heir to the Galuh Kingdom in West Java. The story is intended for readers in years 4, 5, and 6, or roughly for 10- to 12-year-old readers. The story was part of representation study mentioned earlier. We argue that the story, in addition to other stories on the same website, is biased against female characters.

We asked 24 teachers from Medan, Malang, and Ambon to participate in this reception research. The three cities were selected as they represent the western, the central, and the eastern parts of Indonesia respectively. First, we supplied the teachers with the selected story, and we also showed them the website where the story can be accessed publicly. Afterward, we asked them to fill out a questionnaire containing ten questions. The questionnaire contains ten open-ended questions which revolve around characterization and character development. The open-ended questions were chosen in order to provide participants with room for more elaborate critical engagement.

The answers collected from the questionnaire were then analyzed using discourse analysis. We identified key words from answers to each question. We also gathered information on the background of each participant, including their age, gender, location of their workplace. In addition, we also study the contemporary socio-cultural contexts. These pieces of information are important to put the reception in context as we believe that discourse is both constitutive and constituted (Jorgensen & Phillips 2002). Afterwards, we drew patterns and themes. Lastly, we reviewed our results and drew a conclusion. We discuss our findings below.

3 RESULTS AND DISCUSSION

Our analysis shows that these teachers tend to overlook gender biases in the sampled folktale. Regardless of the teachers' gender, age, and cultural origin, they barely noticed the biased characterization and treatment of the male and female protagonists. In doing so, teachers risk sustaining the normalized gender biases, which tend to be disadvantageous to women, and maintaining practices which tend to discriminate against women.

In the first two questions, the research participants were to write four impressions on the character of *Dewi* Samboja and *Pangeran* Anggalarang (Prince Anggalarang), the female and male protagonists of the story respectively. In their impressions of *Dewi* Samboja, we found twenty keywords, with kind, pretty, and obedient to her parents being the most mentioned. As to the impressions of *Pangeran* Anggalarang, the traits which were most mentioned include courageous/brave (both in the sense of fighting his enemies in war and facing his future father-in-law), obedient to the elderly, and handsome.

The stereotypical feminine and masculine traits in the readers' perception are further reinforced in their answers for the third and fourth questions on why the female and male characters choose their life partner. The third question asks the participants reception on the reason Anggalarang chooses Samboja as his future wife. Among the six keywords found, the top three include Samboja's kindness, beauty, and compassion. In question where participants were asked about their perception on why *Dewi* Samboja accepted *Pangeran* Anggalarang's marriage proposal, there are ten keywords found, the topmost answers being related to parental approval, in Anggaralang's courage and bravery in facing Samboja's father to get his approval (19 mentions). The second most popular perception is Anggalarang's being a handsome prince and kind with three mentions each.

The above findings show that the stereotypical characterizations of ideal feminine and masculine traits leave strong imprints in the readers' minds. Being born and raised in Indonesia, we can claim that ideal femininity is often characterized by beauty and kind-heartedness. While beauty is undoubtedly feminine, kind-heartedness may sound neutral. Indeed, male characters can be kind too. Yet, such characterization is not central to ideal masculinity in the story, unlike bravery and courage. Likewise, several research participants also mention *Dewi* Samboja being brave and able to fight, but these traits are not the common impression readers get from the story. Despite the possibility for more critical engagement in answering the reception questions, none of the research participants challenge the stereotypical characterization, which further validates our argument that they are mostly lacking sensitivity towards biased characterization of gender in teaching materials. These findings thus also reinforce our argument (Zahro et al. 2020) that physical beauty remains the anchor in the characterization of female protagonists, and bravery in that of male protagonist, in the written versions of Indonesian folktales published after 2000, where gender equality has been the subject of numerous open public debates, unlike during the authoritarian regime (Robinson 2006).

The fifth, sixth, and seventh questions are related to readers' expectation on Samboja, a female heir to a kingdom. These questions include what skills and training the readers expect Samboja to have learned; whether the story indicates that Samboja has learned those skills; and what they expect Samboja to do with the skills. We found thirty keywords, and the top five include compassion, putting people's welfare first, martial arts, wisdom and strategic thinking. The participants almost unanimously agreed that Samboja has been trained in the mentioned skills. Then, participants expressed that they expected Samboja to continue learning and training, stay true to herself, and get married to a prince; other expectation keywords received a lower number of mentions.

The answers to the set of three questions show that readers support Samboja being a female heir and thus must possess strong leadership skills. Female beauty is not mentioned at all in these questions. However, bias emerges in the further expectation (question 7) which suggest that some of the participants believed that being a female heir, Samboja still needs to be married strategically to a prince who will support her in ruling her kingdom.

The eighth and ninth questions explore the participants' responses towards two incidents, which we deemed highly gendered: Anggalarang's action to save Samboja from the pirates, and Samboja's marriage proposal to Patih Sawunggaling upon the death of her first husband, Anggalarang. In answering the eighth question, the research participants tended to normalize or even valorize Anggalarang's bravery and sacrifice – calling the act as a husband's sacrifice for the sake of his beloved (13 mentions), a husband's responsibility (10 mentions), and a husband's duty to protect his wife. The last two can be merged since the sense we gleaned from the answers is the same – it is a husband's responsibility to protect his wife. The open-ended nature of the question only inspired one participant to question Samboja's martial arts skills despite her being trained. Furthermore, in answering question nine, research participants tend to normalize Samboja's strategic marriage to *Patih* Sawung Galing, his deceased husband's former confidante, because Samboja needs a strong male ally to win her throne back; Samboja needs a man to protect her; and Samboja needs to move on with her life with a new partner after the death of her husband.

The responses to the two questions above clearly indicate internalized patriarchy. The masculine act of sacrificing oneself for his beloved or for his nation is commonly found in other tales of princes and princess around the world. In Western folklore tradition, for instance, many princesses are waiting to be saved by a prince charming in order to be able to live happily ever after. Such narrative normalizes women's helplessness in the face of danger and men's chivalry to offer protection and sacrifice. Moreover, the perception of feminine normalized helplessness is further reinforced in the participants' response towards question nine. These answers imply the participants' perception on women being less capable leaders, who can only be strong if accompanied by strong male allies or husbands.

The last question is a conclusion question. We asked the participants' conclusion on what it took to be a female heir. We found 27 keywords in the collected answers, with the top five being putting the people's welfare ahead of hers, being kind, loving her people, being wise, as well as having compassion and beauty. These answers show that the participants seem to agree that it takes particular soft skills to succeed in becoming a great female leader. Yet, they seem to not leave beauty in the ideal character trait of a female leader.

4 CONCLUSION

Overall, the findings not only reinforce our argument that the story is gender biased, but also reveal that Indonesian teachers who are expected to be more critical in engaging with such stories still normalize the stereotypical characterization of gender in teaching materials. They barely see the biases, which tend to place women as the weaker sex. Such a tenet is further reinforced in their perception of female leaders requiring protection from male allies, who are likely their husbands. In short, most of the teachers in this study perceive it as normal for a female leader to be pretty, strong, smart, but still be physically and skillfully weak compared to men; thus, she still need a man's protection in order to rule and make her country strong.

The implication of our findings is that awareness raising campaigns and activities are still needed. Such ideas can be carried out in the forms of trainings and workshops on gender responsive curriculum. Teachers at all levels must be sensitive of gender issues and able to address responsively to them in their teaching. Secondly, teaching materials are expected to be more gender responsive in order to cultivate gender awareness and nurture responsiveness in education.

Future researchers can benefit from our research and explore teachers' perceptions further on how gender should be represented in teaching materials and curriculum in general. Our early research design was to conduct focus group discussions with teachers in the abovementioned cities. Yet, the design must be modified as the Covid-19 pandemic hit almost every city in Indonesia. As domestic travel restrictions were tightened, we decided to distribute questionnaires to the targeted participants, instead. We did not follow up with a cloud meeting due to time constraints. Thus, future research can pick up from where we left off.

REFERENCES

Frei, S., & Leowinata, S. 2014. *Gender Mainstreaming Toolkit for Teachers and Teacher Educators*. British Columbia: Commonwealth of Learning.

Jørgensen, M., & Phillips, L. 2002. *Discourse Analysis as Theory and Method*. London, Thousand Oaks and New Delhi: SAGE Publications.

Kurnia, N. 2016. *Kisah Dewi Samboja: Cerita Rakyat dari Jawa Barat*. Jakarta: Badan Pengembangan dan Pembinaan Bahasa.

Robinson, K. 2006. Islamic Influences on Indonesian Feminism. *Social Analysis: The International Journal of Anthropology, 50*(1), 171–177.

The Global Partnership for Education and UNGEI. 2016. *Guidance for Developing Gender-Responsive Education Sector Plans*. Washington: The Global Partnership for Education.

UNESCO. 2017. Gender-Responsive Classrooms Need Gender-Sensitive Teachers. 29 May. Retrieved from https://bangkok.unesco.org/content/gender-responsive-classrooms-need-gender-sensitive-teachers

Zahro, A., Eliyanah, E., & Ahmadi, A. 2018. Women and the Indonesian Folktales: Gender Perspective. *International Journal of Humanities and Cultural Studies, 7*(2), 89–99.

Community Empowerment through Research, Innovation and Open Access – Sayono et al (Eds)
© 2021 Copyright the Author(s), ISBN 978-1-032-03819-3

The performance of a library in improving reading culture during the Covid-19 pandemic

A. Asari*, R. Mahdi, D. Widyartono, P.D. Anggari, A. Prasetyawan, K.M. Raharjo,
A.B.N.R. Putra & I.A. Zakaria
Universitas Negeri Malang, Malang, Indonesia

ABSTRACT: The research aims to describe the strategy of the library in improving the reading culture and to find out what factors affect the performance of library services in increasing public interest in reading during a pandemic. The research method used is the descriptive method using surveys and interviews as well as qualitative analysis. Data was obtained through questionnaires, interviews and observations. The research was conducted in the library of state universities in Malang City. The results showed that the performance of library services in improving reading mints was not optimal; it was indicated that there were still user complaints, the attitude of service officers was not fully responsive and information on service programs was not widely known by users.

Keywords: performance, service, library, interest in reading

1 INTRODUCTION

Libraries as public service institutions should be obliged to provide the best service for the community. Therefore, fundamental changes are needed, especially in terms of improving the quality or performance of library services that are more oriented towards community service satisfaction and are responsive to the dynamics of the service environment, meaning trying to provide the best service and evaluate it based on the point of view of library service users (Yi & Cong 2018). So far, the attitudes and behavior of library service providers are more oriented to the needs of institutions and superiors who tend to be served, thus neglecting services to users (Kiran & Diljit 2017).

In this regard, the quality or performance of professional library services needs to be realized, service officers as human resources in the library who are the spearheads in providing services to users are required to provide quality services and are more oriented towards user satisfaction and are more responsive to challenges and opportunities (Murray & Hackathorn 2016). They must not be fixated on routine activities, have the competence to provide fair services and empower the community so that it can create a society that has a high level of literacy (Kotler 2002.).

According to a theoretical perspective, there has been a shift in the public service paradigm from a traditional public administration model to a new public management model, and finally to a new public service model (Piatak & Holt 2020). In the new public service model, public interests are formulated as a result of dialogue from various sources that exist in society and public services must be responsive to various interests and must be non-discriminatory (Subarsono 2005).

According to Sulistyo-Basuki, services in library science can be interpreted as an activity related to providing information by librarians to users (Basuki 2020). Public services in libraries must also be supported by the competencies that librarians must-have. Sudarsono, B. (2020) identifies the competencies that must be owned, namely: (1) knowing the basis of organizing; (2) having and knowing suitable and effective reference interview techniques; (3) knowing printed and digital information resources; (4) having the ability to provide suggestions and directions to users; (5) being

---

*Corresponding author: andi.asari.fs@um.ac.id

DOI 10.1201/9781003189206-10

able to design and implement information retrieval strategies; and (6) able to guide users in the interpretation and evaluation of information.

According to Abbot in Sulistyo Basuki (Sulistyo-Basuki 2004) in simple terms library performance indicators are management tools designed to help library managers determine how good their service performance is. A librarian/service officer is the spearhead in providing quality services to the user community (Basuki 2020). The assessment of the performance of librarians/service officers is one of the most important elements in assessing library performance. This is in line with what was stated by GT. Milkovich and JW Boudreau (Aye & Nusari 2019), who revealed that performance appraisal is a process carried out to assess employee performance while employee performance is defined as a level where employees meet/achieve specified work requirements.

Levine (2014) suggests three concepts that can be used as a reference for measuring the performance of public organizations, namely responsiveness (responsiveness), responsibility (responsibility), and accountability (accountability). Apart from these three performance indicators, it is also common to use other indicators that are more specific according to Mulyadi and Setiawan (Tangkilisan 2005), namely: (1) building customer satisfaction; (2) having employee work productivity; and (3) generating adequate financial returns. Parasuraman, et al. (2001) identified ten main dimensions in assessing the performance/quality of public services, namely reliability, responsiveness, competence, access, politeness, communication, credibility, security, ability to understand customers and physical evidence.

2 METHODS

The research method used was a descriptive research method using surveys, in-depth interviews and qualitative design (Hernon 2008). Descriptive research is intended for careful measurement of certain social phenomena. (Noble & Smith 2015). Descriptive research is a form of research that is most basic, aimed at describing existing phenomena, both natural phenomena and human engineering (Brinkmann, & Kvale 2015).

The variables in this study were the elements of performance measurement of library services, which consist of responsiveness, reliability, service ethics, and service facilities. In this study, data collection was carried out in several ways, including interviews, which were conducted using an interview guide that was focused and flexible. Interviews were needed to deepen various interpretations, perceptions, and perspectives from various policies related to existing discussion topics. Interviews conducted in this study included interviews with library users and library service providers, namely elements of library leaders and library service officers or librarians. In addition to interviews, data collection was strengthened by distributing questionnaires in the form of questions that have been designed in such a way by researchers and addressed to library users with questionable material regarding performance indicators including responsiveness, reliability, service ethics, and service facilities. Then, the results was analyzed by simple descriptive statistics. The final data collection used observation, namely the collection of field data by looking at the performance of library services and the atmosphere of college libraries in Malang City which also affect the performance of library services.

3 RESULTS AND DISCUSSION

Libraries as one of the public service institutions are required to provide optimal performance to improve the reading culture and meet the needs of the user community. With performance appraisal, it can be evaluated to what extent the quality of library service performance in improving the reading culture of the community was. The indicators used to measure the performance of library services include dimensions of reliability. These dimensions are analyzed and interpreted as follows.

3.1 *Procedures and timeliness of service*

Easy service procedures measure the reliability in providing services to library users in supporting the development of a reading culture.

Table 1. Ease of service procedures.

Kind of service	User Complaints				Amount	
	yes		not			
	f	%	f	%	f	%
Member registration	89	90.8	9	9.2	98	100.0
Catalogs and references	67	68.4	31	31.6	98	100.0
Borrowing books	62	63.3	36	36.7	98	100.0
Borrowing newspapers and magazines	69	70.4	29	29.6	98	100.0

Table 2. Timeliness of service.

Kind of service	User Complaints				Amount	
	yes		not			
	f	%	f	%	f	%
Member registration	88	89.8	10	10.2	98	100.0
Catalogs and references	72	73.5	26	26.5	98	100.0
Borrowing books	63	64.3	35	35.7	98	100.0
Borrowing newspapers and magazines	73	74.5	25	25.5	98	100.0

Table 3. Completeness of the collection.

Collection type	Complete collection				Amount	
	yes		not			
	f	%	f	%	f	%
Book	41	41.8	57	58.2	98	100.0
Magazine	44	44.9	54	55.1	98	100.0
Newspaper	68	69.4	30	30.6	98	100.0

However, in reality, service procedures have not fully demonstrated reliability in providing services to users, especially in terms of procedures for borrowing books. As many as 36.7% of users stated that the book borrowing procedure had not made it easy, regarding the lack of convenience in service procedures. The procedures for the retrieval of information both through card catalogs and OPAC catalogs are still considered quite difficult for users.

Timeliness of service is where the implementation of public services can be completed within a predetermined period (KepMenPan 81/1995). Indicators of timeliness of service include timeliness of member registration services, catalogs, and references, borrowing books, borrowing newspapers and borrowing magazines (Table 2).

The survey results show that the punctuality of service has not shown optimal performance. Especially for the book lending service, 35.7% of users still stated that there was no timeliness of service in getting the collection to be borrowed and in catalog and referral services as many as 26% of users stated that the service was not on time. Meanwhile, 35.7% of users stated that the service time for borrowing books was not on time, this is related to the book lending procedure, where users still have to wait a long time to get the books they need so that users do not get certainty of service time.

3.2 Completeness, accuracy, and up-to-date collection

Then the indicators of the completeness of the collection consist of the completeness of the collection of books, magazines, and newspapers/newspapers (Table 3).

Table 4. Collection accuracy.

Collection type	Complete collection				Amount	
	yes		not			
	f	%	f	%	f	%
Book	45	45.9	53	54.1	98	100.0
Magazine	59	60.2	39	39.8	98	100.0
Newspaper	76	77.6	22	22.4	98	100.0

Table 5. Recent collections.

Collection type	Up-to-date collection				Amount	
	yes		not			
	f	%	f	%	f	%
Book	37	37.8	61	62.2	98	100.0
Magazine	59	60.2	39	39.8	98	100.0
Newspaper	79	80.6	19	19.4	98	100.0

The table above shows that most of the users stated that the collections in the library were incomplete. Especially for book collections, as much as 58.2% of users stated that the book collection was incomplete and for magazine collections as much as 55.1% of the respondents stated that it was incomplete. Due to the large number of users who stated that the collections of higher education libraries in Malang City were not complete. An accurate library collection is a reflection that the user's information needs can be met. Collection accuracy indicators include the accuracy of the collection of books, magazines and newspapers (Table 4).

In reality, the existing collections have not fully fulfilled the user's information needs. As many as 54.1% of users stated that the existing book collections were inaccurate, that is, they were not relevant to the user's information needs. This is closely related to the incompleteness of the existing collections. Where the policies in the development of collections are more emphasized in the fields of social sciences and humanities which have not been socialized to users. The current library collection is one of the determining factors in the performance of library services. Where the library must be able to anticipate the information needs of its users, which are dynamic and always changing, following changes in social values in society, through efforts to develop the latest collections, the library can meet the needs of its users. The indicators of up-to-date collections include the latest collections of books, magazines, and newspapers (Table 5).

In terms of the updating of book collections, the majority of users (62.2%) stated that the existing book collections were not up to date. The lack of sophistication of this collection is related to the procurement of library collections that must be guided by government policies through Presidential Decree No. 17 of the year 1980 regarding the system for procurement of goods that must go through vendors and partners. So that the development of library collections is difficult to meet the needs of users for current collections. Not to mention the collection processing process which still takes quite a long time.

3.3 Service personnel communication skills

Indicators of the communication skills of service officers consist of the communication skills of information service officers, member registration, the borrowing of books, and the borrowing of newspapers and magazines, catalogs, and references (Table 6).

In general, the table above shows that most of the users stated that existing service officers had good communication skills. Especially in the case of member registration services, where as many as 86.7% of users stated that service officers had good communication skills. However, there were still quite a lot of users who stated that service officers did not have communication skills, namely

Table 6. Communication skills of library service officers.

Service attendant	Communication skills				Amount	
	yes		not			
	f	%	f	%	f	%
Information service	69	70.4	29	29.6	98	100.0
Member registration	85	86.7	13	13.3	98	100.0
Catalogs and references	66	67.3	32	32.7	98	100.0
Borrowing books	67	68.4	31	31.6	98	100.0
Borrowing newspapers and magazines	62	63.3	36	36.7	98	100.0

as many as 36.7% of users stated that newspaper and magazine loan service officers still did not have good communication skills.

4 CONCLUSION

The development of existing collections has not been able to meet the needs of the user community, especially for the type of book collection. University libraries in Malang City are still late in providing collections, the sophistication of collections in libraries is not the same as the sophistication of collections circulating outside the library. This is related to the collection development system which is still fixated on the rules of the procurement system so that collection development always experiences delays. On the other hand, the time-consuming processing of library materials also plays a role in providing up-to-date collections.

REFERENCES

Aye, T., Ameen, A., & Nusari, M. 2019. Factors influencing job performance of employees from international non-profit organizations in Myanmar. *International Journal on Recent Trends in Business and Tourism (IJRTBT)*, 3(2), 56–68.
Basuki, S. 2020. Profesi Dan Konsep Pustakawan Dalam Konteks Indonesia. *Media Pustakawan*, 17(1&2), 75-83.
Brinkmann, S., & Kvale, S. 2015. *Interviews: Learning the craft of qualitative research interviewing* (Vol. 3). Thousand Oaks, CA: Sage.
Hernon, P., & Schwartz, C. 2008. Leadership: Developing a research agenda for academic libraries. *Library & Information Science Research*, 30(4), 243–249.
Kiran, K., & Diljit, S. 2017. Antecedents of customer loyalty: Does service quality suffice?. *Malaysian Journal of Library & Information Science*, 16(2), 95–113.
Kotler, Philip. 2002. *Manajemen Pemasaran*. Jakarta : Prehaliilindo.
Levine, S. S., & Prietula, M. J. 2014. Open collaboration for innovation: Principles and performance. *Organization Science*, 25(5), 1414–1433.
Lubis, Ella H. 1999. Penilaian kinerja individu (performance Appraisal). *Jurnal Usahawan* No. 11 TH XXVIII. (34–36).
Murray, A., Ireland, A., & Hackathorn, J. 2016. The value of academic libraries: Library services as a predictor of student retention. *College & Research Libraries*, 77(5), 631–642.
Noble, H., & Smith, J. 2015. Issues of validity and reliability in qualitative research. *Evidence-based nursing*, 18(2), 34–35.
Piatak, J. S., & Holt, S. B. 2020. Prosocial behaviors: A matter of altruism or public service motivation?. *Journal of Public Administration Research and Theory*, 30(3), 504–518.
Subarsono, A. G. 2005. *Analisis kebijakan publik: konsep, teori dan aplikasi* (Vol. 138). Pustaka Pelajar.
Sulistyo-Basuki. 2004. *Pengantar Dokumentasi*. Bandung: Rekayasa Sain.
Sudarsono, B. 2020. Pustakawan dan Perpustakaan dalam menghadapi tantangan di Era Global. *Media Pustakawan*, 18(3), 1–8.
Tangkilisan, Hessel Hogi S. 2005. *Manajemen publik*. Jakarta: Grasindo.
Yi, K., Chen, T., & Cong, G. 2018. Library personalized recommendation service method based on improved association rules. *Library Hi Tech*.

Community Empowerment through Research, Innovation and Open Access – Sayono et al (Eds)
© 2021 Copyright the Author(s), ISBN 978-1-032-03819-3

The development of instructional materials for reading literacy using the WISE approach

E.T. Priyatni*, A.R. As'ari, Nurchasanah & Suharyadi
Universitas Negeri Malang, Malang, Indonesia

ABSTRACT: This study aimed to develop instructional materials for reading literacy using the WISE approach. WISE stands for wondering, investigating, synthesizing, and expressing. The development of the instructional materials followed ADDIE stages (analysis, design, development, implementation and evaluation). The development of materials for reading literacy using the WISE approach consists of seven chapters: introduction, WISE principles, activities to promote the wondering skill, activities to promote the investigating habit, activities to promote the synthesizing skill, activities to promote the expressing skill and follow-up activities. The formation of the WISE technique in students is possible if wondering is encouraged in students. Students also need to stimulate curiosity by themselves to develop and implement an investigation design, analyze the investigation results and use creativity to synthesize and express ideas.

Keywords: instructional materials, reading literacy, wondering, investigating, synthesizing, and expressing

1   INTRODUCTION

One of the primary concerns of education in Indonesia is students' poor literacy. The result of PISA in 2018 showed a decrease in Indonesian student literacy scores compared to the 2015 PISA result (OECD 2019). The data further revealed that Indonesian students scored 371 in reading, 379 in mathematics and 396 in science. A significant decrease was also observed in the students' reading literacy score (OECD 2019). There was not a single Indonesian student who was able to achieve a level-5 reading literacy score. In fact, only 0.4% of the students were able to perform level-4 literacy skills, while the rest scored between levels 1-3 in terms of reading literacy. The students' low interest in reading was also proven by 2012 UNESCO statistical data, suggesting only 1 (0.1%) out of 1000 people in Indonesia have a good reading interest (OECD 2014). Improving the quality of reading literacy in Indonesia is an important task in national education (Priyatni & Nurhadi 2017).

The low reading literacy is not proportional to the abundance of information that students can access every day through technology. This information comes in the form of texts, images, tables, audio, and audio-visual components. The abundance of information that is present through technology is irreversible. Every minute, students can read articles or statuses posted on social media. Not all information is true and accurate. Therefore, students need to develop high literacy skills to filter, select and sort information and hoaxes. Literacy is the heart of human life and must be mastered and reached by everyone (Narey 2017). Everyone must be literate. Everyone must have basic literacy, including reading literacy.

The notion of literacy continues to develop, along with the development of digital technology. In the beginning, literacy was used to describe people's ability to decode and encode texts (Gurak 2001). However, since the mid-20th century, the concept of literacy has changed to literacy that

---

*Corresponding author: endah.tri.fs@um.ac.id

DOI 10.1201/9781003189206-11

focuses primarily on reading and writing, the two skills which form the basis for literacy in various matters (Kalantzis & Cope 2015). Literacy also includes reading, writing, and numerical literacy, which constitute three basic life skills (Kalantzis & Cope 2015). In the next development stage, literacy is defined as knowledge literacy, an understanding of something (knowledge). Furthermore, literacy is defined as the ability to think critically-creatively, which is supported by the tradition of reading-writing-numerical literacy and figures (pictures/tables) literacy. Both thinking activities have a different focus: critical reading is focused on assessing, while creative reading is focused on the process of making or producing (Paul & Elder 2008; Priyatni & Martutik 2020). This is in line with the 2015 PISA, which defines reading literacy as the ability to understand, apply, evaluate and reflect on texts to achieve goals, expand knowledge, and participate actively in social life (Csapó & Funke 2017; OECD 2019).

Research on reading literacy has been carried out in various contexts in the world, including in Indonesia. Wu and Peng (2017) examined the effect of modality on students' reading literacy. They found that students had better reading literacy through printed reading materials. Another study by Tan (2015) showed that the existence of Wikis, email, blogs and other online reading sources could improve students' reading literacy. Previous research carried out by Boyle et al. (2019) suggested that reading literacy is also important for students with special needs. The School Literacy Movement (SLM) carried out in schools in Indonesia can increase student interest in reading and reading activity frequency (Sutrisna et al. 2019). The results of the preliminary study indicate that the School Literacy Movement program has not been able to make students literate in reading nor build students' ability to reason. To increase the role of SLM in promoting reading literacy, research and development of a WISE-oriented integrative reading literacy prototype was carried out simultaneously.

WISE stands for Wondering, Investigating, Synthesizing and Expressing. WISE is not a learning model; WISE is a goal. By carrying out WISE-oriented learning, students will have: (1) a good wondering habit, that is, asking a lot of questions and always wanting to know more, (2) strong investigating skills, which means being willing and able to find the necessary information, (3) solid synthesizing skills, namely being able to link and draw conclusions, and (4) increased engagement due to having a purpose and being skilled in communicating ideas or findings.

2 METHODS

This study employed a Research and Development (R & D) method to develop a product. The product developed in this study was the prototype of WISE-oriented instructional materials for reading literacy. The development of the prototype followed the stages of ADDIE (analysis, design, development, implementation, and evaluation (Spector et al. 2014). The analysis of the initial conditions and the requirements needed for developing the prototype was conducted at the Analysis stage. This stage was followed by designing and developing the prototype. Then, learning and materials experts, as well as practitioners (teachers), were invited to examine the validity of the prototype. Inputs from the experts and the practitioners were used to revise the product. The revised product was implemented in the classroom, and the implementation of the product was observed, evaluated, and reviewed.

This article focuses on describing the process of developing the prototype and the results of expert validation rather than elaborating on the implementation of the product to potential users/students. The data of this study was collected in the form of comments, critiques, suggestions, and judgments from the experts and practitioners team. The learning and material experts hold doctoral degrees and have more than ten years of teaching experience. The practitioners involved for expert validation were the teachers from SMPN 1 Tana Tidung Regency in North Kalimantan. The researchers in this study acted as the key instruments in data collection, data analysis, and data interpretation assisted by an instrument called product review guidelines. Data analysis was focused on analyzing data from the experts/practitioners involved in expert validation. The results of this analysis were then used to improve the product.

3 RESULTS AND DISCUSSION

The products developed in this study were reading literacy instructional materials with WISE orientation. WISE is not a learning model but a goal. It means that when these learning materials are used in learning, students' wondering, investigating, synthesizing, and expressing skills can be promoted.

The prototype of the materials consists of seven chapters: (1) introduction, (2) WISE principles, (3) activities to promote the "wondering" skill, (4) activities to promote the "investigating" habit, (5) activities to promote the "synthesizing" skill, (6) activities to promote the "expressing" skill and (7) follow-up activities. The introduction section presents the context, focus and method of the study. In Chapter 2, the nature or the essence of WISE is presented, including the definition, purpose, and elements of WISE. Chapters 3, 4, 5 and 6 contain activities to cultivate wondering, investigating, synthesizing, and expressing skills, respectively, which are described below.

In detail, the activities to promote Wondering section present a multimodal text stimulus, which can be in the form of images, tables, infographics, diagrams, or continuous texts (in the form of sentences). The stimulus chosen is attractive and has the elements of novelty and local wisdom to arouse students' curiosity. From this stimulus, students are asked to ask questions. The next section is concerned with activities to promote the investigation habit. The main aim of an investigation activity is to investigate the major ideas found in the activity of wondering through the research process based on facts and by recording the results of the investigation accurately. The investigation process includes three activities, namely, planning, implementing, and processing. Activities at the planning stage include selecting which questions will be followed up (answered or investigated), finding out what information should be collected, where the information sources are, how the information is obtained, what tools/instruments are needed to gather the information. At the implementation stage, the activities carried out by students include interviewing, conducting surveys, surfing in cyberspace, and conducting safe experiments. At the processing stage, the activities carried out by students are assessing the validity of the data, storing the data safely, compiling the data regularly, and presenting the data properly.

The investigation process includes three activities: planning, implementing, and processing. Activities at the planning stage include selecting which questions will be followed up (answered or investigated), finding out what information should be collected, where the information sources are, how the information is obtained, what tools/instruments are needed to gather the information. At the implementation stage, the activities carried out by students include: interviewing, conducting surveys, surfing in cyberspace, and conducting safe experiments. At the processing stage, the activities carried out by students are assessing the validity of the data, storing the data safely, compiling the data regularly, and presenting the data properly.

Synthesizing activities primarily aim to synthesize big ideas developed through investigation, link the data/information obtained from the investigation and determine the relationship between the ideas. Activities undertaken by students to hone their synthesizing skill include checking the validity of the data sources, checking the quality and credibility of the data sources, comparing the data sources with other data sources, checking whether there are contradictory data sources, linking the data with other data, drawing conclusions and compiling findings.

The main purpose of expressing activities is to communicate big ideas that have been developed, communicate new ideas, use an appropriate format, communicate new ideas that have been developed clearly and creatively, and communicate new ideas using various modes and technologies. Expressions must be thought through, and they must not be careless. Students are encouraged to present new ideas that have been found in the form of videos, audios, infographics, full texts that contain tables, diagrams, etc.

The product prototype was validated by two learning material experts and two learning experts. The researchers presented the developed product via virtual meeting, and then the experts were asked to provide comments on the product. The summary of the results of the expert's analysis can be seen in Table 1.

Table 1. The expert validation results.

No.	Analyzed Aspects	Analysis Results
1.	Stimulus to stimulate the development of wondering skill	Mostly, the stimulus is provided in the form of images and infographics. The variation of stimulus can be full story texts, videos or audio.
2.	Activities to promote the wondering skill	The question column is provided for students because questions should arise from students so that they have a more positive impact. If students have not asked wondering questions or questions that arouse curiosity and require investigation, the teacher can guide students by giving guiding questions.
3.	Activities to promote the investigation skill	If there is more than one wondering question, let students choose which question to answer through the investigation. Generally, investigative activities have been provided by the teacher. Let students plan ways to investigate, carry out, and process the results. Investigations work best when students are left to figure out how to investigate. Let the students find a way to find answers to the selected questions, to whom they should ask to find answers, what questions to ask.
4.	Activities to promote the synthesizing skill	Synthesizing information is not easy. Add examples of finding data — then invite students to classify, determine which one is better, or make conclusions. Provide a column that guides students to collect facts, then ask students to validate which one is the best, which one is chosen for what.
5.	Activities to promote the expressing skill	The most important thing in expressing activities is that students can communicate their ideas creatively. Students can organize their ideas attractively by using various modes (oral writing, audio, audio-visual, infographics, graphics, tables, etc.).

The practitioners (teachers) invited also provided suggestions on the product developed in this study. The teachers' suggestions can be seen in Table 2.

Based on the input from material experts, learning experts and practitioners (teachers), the product was revised from the aspect of the stimulus which is used to foster students' wondering skill by adding various stimuli, which are familiar to the students, and exploring local wisdom in the area where the students live. This is in line with the direction of reading literacy, where the focus of the study is text (OECD 2018), and text is a meaningful message (Anderson & Anderson 2003) and is multimodal (Kalantzis & Cope 2015). Literacy in the modern era must be directed to understand multiliteracies, namely multi-contextual and multimodal texts (Kalantzis & Cope 2015). Multi-contextual texts include texts relating to community setting, social role, interpersonal relations, identity and subject matter, while the multimodal scope includes written, visual, spatial, tactile, gestural, audio, and oral (Kalantzis & Cope 2015).

Suggestions for incorporating local cultural values into WISE-oriented reading literacy are also used as a reference for revising the product. This is in line with the policy of strengthening character education stated in the 2013 curriculum and contained in the Merdeka Belajar curriculum's grand design. Local wisdom is the basic asset given to students to familiarize them with existing noble values, either as individuals, communities, or citizens. Local culture has high moral values that can be used as a source to strengthen student character. Character is a set of moral beliefs or personality which is formed from the internalization of various virtues that are believed and used as a basis for shaping a point of view, thinking, behaving and acting (Stevenson 2006). Character education

Table 2. Suggestions from the practitioners.

No.	Analyzed Aspects	Analysis Results
1.	Stimulus to stimulate the development of the wondering skill	The stimulus should be familiar to the students. The stimulus should take advantage of local wisdom so that students can get to know the richness of culture, nature and traditions that exist in the area. The stimulus should be associated with the problems that exist around the students so that they get used to solving problems around them. Stories can also be used as a stimulus, not necessarily in the form of pictures.
2.	Activities to promote the wondering skill	Lead questions are needed so that students can answer wondering questions. Many examples are needed so that students can answer wondering questions. High-level questions should be added.
3.	Activities to promote the investigation skill	How to find information, is also a form of investigation. Investigations do not have to be experiments.
4.	Activities to promote the synthesizing skill	Concrete examples are needed to practice the synthesizing skill.
5.	Activities to promote the expressing skill	There is an activity of giving feedback after students express their work. Feedback can come from netizens - by uploading student work to Youtube.

needs to be continuously conducted, and sources of value for strengthening character should be taken from the nation's noble values and the Indonesian nation's culture. Character education aims to foster the younger generation's personality and form individuals who have good character and can serve as good citizens (Priyatni 2013).

The examples of wondering questions that stimulate critical thinking, creativity, collaboration and investigative communication skills should be added. This suggestion is very important and must be used as the "spirit" to develop WISE-based reading literacy instructional materials. Wondering is essentially an innate skill that waits for stimuli/triggers to grow and develop (L'Ecuyer 2014). Wondering skills must be cultivated from within students. Let the students take the initiative. The teacher is allowed to provide triggers at an early stage only, and then slowly, the teacher allows the curiosity to grow from within the student. This needs to be given special attention because the initiative that always comes from the teacher will numb the curiosity of the students (L'Ecuyer 2014). When wondering is not grown in students, there will be a mechanical process in learning, so that learning is only repetitive, becoming a routine that is deadly and alienating (L'Ecuyer 2014).

The focus of developing WISE-oriented instructional materials is promoting wondering, investigating, synthesizing, and expressing skills in students. These four goals ultimately lead to the growth of creative thinking skills. Creative thinking is one of the important goals of education (Wolska & Dlugosz 2015). A key role in developing the creativity of students at all ages is played by formal education and teachers are responsible for stimulating students' creative abilities, forming student personality and attitudes that are conducive to creativity, and teaching students creative thinking skills and creative problem solving (Wolska & Dlugosz 2015).

Communication is also an important skill that becomes the focus of developing WISE-oriented reading literacy instructional materials. Expressing requires students to be able to choose different patterns of expression according to the context (Kalantzis & Cope 2015). Cope exemplifies that the way a doctor delivers news about laboratory results to patients will be different from that used when conveying it to the patient's parents (Kalantzis & Cope 2015). Delivering the same message to people of different ages, different social status (student-teacher) requires different speech forms. Expressing is an important skill that allows communication to be established well, smoothly, without any emotional obstacles.

4 CONCLUSIONS

The reading literacy product prototype with wondering, investigating, synthesizing, and expressing (WISE) was developed to promote student curiosity, investigative habits, synthesize skill and ability to communicate ideas appropriately and attractively. To develop the student wondering skill, the stimulus is presented in the form of multimodal texts, which can be in the form of images, tables, infographics, diagrams, or continuous texts (in the form of sentences). The stimulus chosen is an interesting, attractive stimulus and has novelty and local wisdom elements. This curiosity is continued by encouraging students to investigate in their own way. Investigative data is grouped, compared, synthesized and then communicated using various modes. WISE will be formed in students if curiosity is encouraged from within the student, meaning that the students develop curiosity by themselves, design, implement and process the results of investigations, as well as synthesize and express ideas using their creativity.

ACKNOWLEDGEMENT

We would like to thank the Institute for Research and Community Service, Universitas Negeri Malang, for funding this study with the 2020 PNBP grant (contract number: 4.3.590 '/UN32.14.1/LT/2020).

REFERENCES

Anderson, M., & Anderson, K. 2003. *Type Text in English*. Macmillan.

Boyle, S. A., Boyle, D., & Chapin, S. E. 2019. Effects of Shared Reading on the Early Language and Literacy Skills of Children with Autism Spectrum Disorders: A Systematic Review. *Sage*.

Gurak, L. J. 2001. *Cyberliteracy: Navigating the Internet with Awareness*. Yale University Press.

Kalantzis, M., & Cope, B. 2015. *Literacies*. Cambridge University Press.

L'Ecuyer, C. 2014. The Wonder Approach to Learning. *Hypothesis and Theory Article*, 8(October), 1–8.

Narey, M. J. 2017. *Multimodal Perspectives of Language, Literacy, and Learning in Early Childhood. The Creative and Critical "Art" of Making Meaning* (12th ed.). Springer.

OECD. 2014. *PISA 2012 Results: Creative Problem Solving: Students' Skills in Tackling Real-Life Problems: Vol. V*. oecd publishing.

OECD. 2018. *PISA 2018 Reading Literacy Framework*.

OECD. 2019. *PISA 2018 results (Volume I): What 15-year-old Student in Indonesia Know and Can Do. Indonesia-Country Note-PISA 2018 Result*.

Paul, R., & Elder, L. 2008. The Nature and Functions of Critical & Creative Thinking. *Nature, 2008*, 1–52.

Priyatni, E. T. 2013. Internalisasi Karakter Percaya Diri dengan Teknik Scaffolding. *Pendidikan Karakter*, 3(2), 164–173.

Priyatni, E. T., & Martutik, M. 2020 The Development of a Critical-Creative Reading Assessment Based on Problem Solving. *Sage Open*, 10(2), 1–9.

Priyatni, E. T., & Nurhadi. 2017. *Membaca Kritis dan Literasi Kritis* (kesatu). TSmart.

Spector, M., Merrill, D., Elen, J., Bishop, M. J., & Savelyeva, T. 2014. *and Technology*.

Stevenson, N. 2006. *Young Person's Character Education Handbook* (S. Pines (ed.)). JIST Publishing.

Sutrisna, I., Sriwulan, S., & Nugraha, V. 2019. Pengaruh Gerakan Literasi dalam Meningkatkan Minat Baca Siswa. *Parole (Jurnal Pendidikan Bahasa Dan Sastra Indonesia)*, 2(4), 521–528. https://journal.ikipsiliwangi.ac.id/index.php/parole/article/view/2878

Tan, A. 2015. *Creativity in the Twenty First Century Creativity, Culture*. Springer.

Wolska, M., & Dlugosz. 2015. Stimulating the Development of Creativity and Passion in Children and Teenagers in Family and School Environment - Inhibitors and Opportunities to Overcome Them. *Procedia - Social and Behavioral Sciences*, 174, 2905–2911.

Wu, J. Y., & Peng, Y. C. 2017. The Modality Effect on Reading Literacy: Perspectives Students' Online Reading Habits, Cognitive and Metacognitive Strategies, and Web Navigation Skills Across Regions. *Interactive Learning Environments*, 25, 859–876.

Community Empowerment through Research, Innovation and Open Access – Sayono et al (Eds)
© 2021 Copyright the Author(s), ISBN 978-1-032-03819-3

Achievements and barriers to river school community in the conservation of the Brantas watershed

A. Tanjung*, A. Kodir, S. Zubaidah, M. Syaifulloh & S.P. Yunita
Universitas Negeri Malang, Malang, Indonesia

ABSTRACT: Pollution in the Brantas River has a serious impact on reducing its water quality. This problem raises concerns from various groups, including educational institutions. Water Quality Monitoring Community Networks, called by JKPKA is a voluntary organization consisting of schools in the Brantas and Bengawan Solo watersheds committed to preserving the river environment. This study aims to determine the achievements and obstacles of JKPKA in an effort to preserve the Brantas watershed. This study used qualitative research methods. The data collection process was carried out through interviews with the JKPKA central and regional coordinators and several high schools (SMA) teachers who played an active role in preserving the Brantas watershed. We describe the results of this study indicate that the JKPKA achievements include (1) the discovery of the Water Inquiry learning model, (2) recording the MURI record for monitoring river water quality, and (3) contributing to the improvement of the Brantas watershed environment. Meanwhile, the obstacles faced by JKPKA include (1) no support from the local Education and Culture Office, (2) the ineffective succession of management in schools and, (3) dependence on the role of teachers and students in spreading the idea of river schools.

Keywords: River school, achievements, barriers, Brantas, watershed

1   INTRODUCTION

The condition of the Brantas River increasingly concerns due to domestic and industrial waste pollution. One of the serious impacts arising from this pollution is a decrease in water quality as a result of population density and economic activity (Kodir 2019). About 60% of the waste polluting the Brantas River comes from households. The rest is an industrial waste of toxic and hazardous materials (Widianto 2019). The 2017 Environmental Management Quality Index (IKPLH) report shows that the water quality of the Brantas River is in the alert category, with class 3 quality being polluted (DLH Jatim 2017). These findings indicate that Brantas river water is not suitable for drinking and agriculture. However, if there are no other options, it is necessary to process these two needs. In addition, currently, the Brantas River is also known as a strategic waste disposal site with a composition of 42% diaper waste, 37% plastic, and 30% household waste (Wulandari & Suwanda 2019).

The area most affected by waste pollution is the downstream part of the Brantas watershed, to be precise, between Mojokerto and Surabaya. Measurement of river water quality carried out by ECOTON together with the Kali Surabaya Women's Paralegal (PPKS) in April 2020 shows that the water quality of the Brantas River is getting worse towards the downstream area. Dissolved oxygen content (KOT) in seven locations including Mlirip, Wringinom, Sumengko, Karangpilang, Joyoboyo, Monkasel, and Petekan showed that the consecutive decreasing numbers from Mlirip to Petekan were 4.7 ppm, 3.25 ppm, 3.34 ppm, 1.69 ppm, 2.51 ppm, 1.20 ppm. In the Mlirip

---

*Corresponding author: ardyanto.tanjung.fis@um.ac.id

Mojokerto area, the KOT is still above the standard, namely 4.7 ppm (the KOT standard for class 2 rivers is 4 ppm). In the Wringinom to Petekan area, the KOT is below the standard.

The environmental conservation efforts of the Brantas watershed are the responsibility of all parties. This principle is the background of the formation of JKPKA as a school community that cares about rivers. JKPKA was formed on June 24, 1997, in Malang as a result of a collaboration between IKIP Malang (now Universitas Negeri Malang) and Perum Jasa Tirta 1 Malang involving high school teachers along the Brantas watershed. Until 2019, JKPKA had joined 220 schools from elementary to high school levels located along the Brantas and Bengawan Solo watersheds as members. At first, JKPKA was present as a promoter of river-based learning. However, in its development, the activity program initiated by JKPKA has also contributed to the conservation efforts of the Brantas watershed. The activities include joint monitoring of river water quality, training of teachers and students on environmental education, and scientific meetings.

2 METHODS

This study used qualitative research methods. This research focuses on three segments of the Brantas watershed, namely the upstream, middle, and downstream parts. The data collection process was carried out through several stages, namely direct interviews, observation, and literature study. The interview process was carried out in-depth and openly using the purposive sampling technique. Some of the informants that will be interviewed are the JKPKA regional and central Coordinators. Meanwhile, a focused discussion process was conducted with several high school teachers who took an active role in preserving the Brantas watershed. The data analysis in this study used thematic analysis. This analysis was carried out through several stages (Bryman 2016). First, reviewing the interview transcripts. Afterward, coding the interview transcripts from several quotes and classifying the results of the interviews were based on the topics discussed. And the last one is to interpret findings from predetermined ideas.

3 RESULTS AND DISCUSSION

3.1 *JKPKA programs*

JKPKA is the driving force that coordinates schools to help conserve water resources in the Brantas and Bengawan Solo watersheds. As an organization, JKPKA has the vision to create schools that care about water resources, rivers, and the surrounding environment. Among the work programs formulated to achieve this vision are (1) teacher training, (2) student training, (3) joint monitoring of water quality, (4) Tirta Bhakti campsite and, (5) scientific meetings. The source of funding for the JKPKA work program is taken from the environmental conservation budget of Perum Jasa Tirta 1 Malang as one of the initiators and a BUMN that has responsibility and authority regarding the conservation of the Brantas watershed.

The training provided by JKPKA to teachers is varied. There is training to strengthen research, mental training, and insight into the water environment. The main objective of training for teachers is to enrich knowledge regarding river environment-based learning (Wulandari & Suwanda 2019). All training events are carried out centrally in the tourist park owned by Perum Jasa Tirta. As a follow-up, a teacher who has attended training from JKPKA is obliged to train students in their respective schools. Training for students can be carried out through internalization in the subject matter or through student activities in the form of extracurricular activities. Suppose the existence of JKPKA has been institutionalized in the school. In that case, the student or schoolteacher representatives have the right to participate in joint monitoring of water quality, water quality camps, and scientific meetings between JKPKA members.

Joint monitoring of water quality is JKPKA's flagship work program as well as the basis for naming this community. Water quality monitoring by JKPKA applies the biotilic method by involving students and guidance teachers in each school going directly to the river. In its development, the consistency of this monitoring has resulted in a learning model called water inquiry. Another case with water quality monitoring, which only takes place in one day, the Tirta Bhakti Camping Activity Program (PEKERTI) is held for two days and one night. In fact, this camp is an innovation from the Saturday-Sunday (PERSAMI) camp, which is synonymous with the Boy Scout movement activity, which is now a compulsory extracurricular activity in schools at all levels. PEKERTI is designed by determining the camping location adjacent to the Brantas or Bengawan Solo rivers. Thus, campers can carry out river tracing activities and map the river environment to be evaluated and followed up in reforestation actions.

The last JKPKA program is a scientific meeting. The scientific meeting is the highlight of the JKPKA work program series. This program is a forum for creating and appreciating school students who are members of the JKPKA membership. At the scientific meeting event, there are various types of competitions such as river ambassadors, environmental essays, scientific works, etc. On the other hand, this program also intends to popularize the existence of JKPKA in the minds of students. They are the potential generation to maintain river water quality in the present and the future. Therefore, to develop environmental awareness and skills, it is necessary to know the importance of maintaining the conservation of water resources' quality. A Water Quality Monitoring Communication Network (JKPKA) was established. JKPKA is indeed not the only movement concerned with conserving water resources in Indonesia, but only JKPKA is coordinating schools around the Brantas river to conserve water resources.

3.2 *JKPKA achievements*

For 23 years contributing to preserving the Brantas watershed JKPKA has made several achievements. First, the discovery of the water inquiry learning model by Soetarno Said as the JKPKA Central Coordinator. The creation of this learning model is the fruit of JKPKA's commitment to periodically monitoring the water quality of the Brantas River. Water inquiry is actually a learning model that focuses on fulfilling students' curiosity through real learning media, namely the Brantas River. This model has been officially recognized and announced by Prof. Dr. Soeparno (UM rector for 2006–2014) in the 2014 JKPKA scientific meeting. As with other scientific learning models, the water inquiry learning model also has a learning syntax. The learning syntax of this model includes (1) forming student groups (grouping), (2) data collection (observation), (3) data analysis (analyzing), (4) group discussion (discussing) and, (5) delivering the results of the discussion (communicating).

JKPKA's second achievement was successfully holding records at the Indonesian Record Museum (MURI) concerning river water quality monitoring activities. The recording of this record took place on April 22, 2017, carried out simultaneously in two rivers, namely Brantas and Bengawan Solo, both upstream, middle, and downstream. The MURI record-breaking involved more than 1,500 students and teachers (SD, SMP, SMA, and equivalent) who were JKPKA members from Kota Batu, Malang, Blitar, Tulungagung, Kediri, Jombang, Gresik, Surabaya, Madiun, and Boyolali.

JKPKA's third achievement is to contribute to improving the environment of the Brantas River. The improvement effort is in the form of a campaign to care for the environment by students of SMAN 2 Malang (JKPKA members upstream of the Brantas River). This campaign is held regularly every week at the Car Free Day (CFD) event located in Ijen Street, Malang City and has been carried out since 2018 until now. In addition, there is environmental advocacy driven by students of SMAN 21 Surabaya. Based on the narrative of Drs. Budi Santoso M.Pd, as the coach of JKPKA SMAN 21 Surabaya, his students had reported a company dumping garbage in the Brantas River. Then the company committed acts of intimidation against his party. This happened because companies reportedly did not have an Environment Impact Assessment (EIA). To date, thanks to the reporting, they have an EIA document.

3.3 JKPKA barriers

During the 23 years of the process of empowering schools in the Brantas watershed, there have also been several obstacles that have slowed down the development of JKPKA, both external and internal. JKPKA's external obstacle is the lack of significant support from the local Education and Culture Office (provincial and municipal levels). The lack of support from the Education Office has made the existence of JKPKA in school institutions very dependent on leadership policies (Fujimoto 2013). In fact, the position of school principal will continue to experience changes within a certain period, likewise, with the policy.

The second obstacle to JKPKA is the ineffective succession of management in schools. Like other voluntary organizations, the transformation of JKPKA management in schools is also prone to failure. The reason is, so far in the operational management of JKPKA there has been no structural succession planning. In fact, it is common in the teaching profession to transfer positions or places of work (schools) (Febriana et al. 2016). In addition, a single management factor also hinders the transformation process of JKPKA coaching in schools. Succession problems will arise when JKPKA supervisors enter their full term. As a result, the succession process was only based on appointment alone without any prior regeneration.

The third obstacle of JKPKA is the dependence on the role of mentor teachers and students in each school to ground the idea of schools that care about the water environment. The absence of an organizational umbrella from the Education and Culture Office has resulted in misperceptions about a water school's idea among teachers (Fujimoto 2013). Then, the assumption that the topic of river schools is only for students majoring in mathematics and natural sciences (MIPA) -especially biology at the high school level also hinders the distribution of this idea to other specializations. As a result, the dissemination of ideas born from JKPKA requires a complete commitment and a proficient communication strategy from teachers and students who have been trained in the JKPKA training program.

4 CONCLUSION

To sum up, JKPKA is a river school community committed to preserving the environment in the Brantas and Bengawan Solo watersheds. The JKPKA work program consists of (1) teacher training, (2) student training, (3) joint monitoring of water quality, (4) Tirta Bhakti Camp and, (5) scientific meetings. The achievements of JKPKA for 23 years of actively preserving the Brantas watershed include (1) the discovery of the Water Inquiry learning model, (2) recording the MURI record for monitoring river water quality, and (3) contributing to the improvement of the Brantas watershed environment. The obstacles experienced by JKPKA include (1) no support from the local Education and Culture Office, (2) ineffective succession of management in schools and, (3) dependence on the role of teachers and students in spreading river school ideas. The implication of our study derives from our finding on the uniqueness of the knowledge and information work carried out by JKPKA.

REFERENCES

Bryman, A. 2016. Social Research Methods. 5th ed. Oxford: Oxford University Press Inc.

Darmawan, B., Saam, Z., & Zulkarnaini. 2010. Hubungan Pengetahuan, Sikap, Perilaku dan Peranserta dengan Kesadaran Lingkungan Hidup Serta Kesanggupan Membayar Masyarakat Sekitar Bantaran Sungai Di Kota Pekanbaru. *Journal of Environmental Science*, 2 (4) : 103–116. ISSN 1978-5283.

Dinas Lingkungan Hidup Jawa Timur. 2017. Dokumen informasi Kinerja Pengelolaan Lingkungan Hidup Daerah Provinsi Jawa Timur.

Ecoton.or.id. 2020. Jelang PSBB Kualitas Kali Surabaya dan Kali Mas di Bawah Baku Mutu Kelas Air. [Cited 2020 July 21]. Avaible from: http://ecoton.or.id/2020/04/24/jelang-psbb-kualitas-kali-surabaya-dan-kali-mas-di-bawah-baku-mutu-kelas-air/

Febriana, I., Ibrohim, & Mahanal, S. 2016. Potensi Pembelajaran Inkuiri dalam Menumbuhkan Sikap Siswa terhadap Lingkungan. *Prosiding*. Semnas Pend IPA Pascasarjana UM, Vol. 1, 2016, ISBN: 978-602-9286-21-2.https://www.menlhk.go.id/site/single_post/2611/434-sekolah-raih-penghargaan-adiwiyata-tahun-2019

Fujimoto K. 2013 *Brantas River Basin Development Plan of Indonesia*. In: Nissanke M., Shimomura Y. (eds) Aid as Handmaiden for the Development of Institutions. Palgrave Macmillan, London

Kodir, A. 2019. Political Ecology of a spring: People's Resistance to the Construction of a Hotel, *Journal of asian sociology* 48 (2), 179–198.

Neolaka, A. 2008. *Kesadaran Lingkungan*. Jakarta: PT. Rineka Cipta

Rosmini, D., Septiono, M A., Putri, N. E., Shabrina, H. M., Salami, I. R. S., & Ariesyady, H. D. 2018. River Water Pollution Condition in Upper Part of Brantas River and Bengawan Solo River. *IOP Conf. Series: Earth and Environmental Science* 106 2018 012059. https://malang.merdeka.com/kabar-malang/dua-pelajar-sma-di-jatim-dinobatkan-sebagai-duta-sungai-2017-170831x.html

Sari, R. A., Prayogo, T. B., &Yuliani, E. 2017. Studi Penentuan Status Mutu Air di Sungai Brantas Bagian Hilir Untuk Keperluan Air Baku. Brawijaya University, Engineering Faculty. Retrived April 2, 2018 from http:// pengairan.ub.ac.id/wp-content/uploads/2017/01/Studi- Penentuan-Status-Mutu-Air-Di-Sungai-Brantas-Bagian-Hilir-Untuk-Keperluan-Air-Baku-Rani- Anjar-Sari-125060400111052.pdf.

Wang, Q., & Yang, Z. 2016. Industrial Water Pollution, Water Environment Treatment, and Health Risk in China. *Environmental Pollution*, 218: 358–365.

Widianto, Eko. 2019. Sungai Brantas Makin Memprihatinkan. [Cited 2020 July 21]. Avaible from: https://www.mongabay.co.id/2019/05/12/sungai-brantas-makin-memprihatinkan/

Wulandari, D., Suwanda, I.M. 2019. Peran Yayasan Ecoton dalam Menumbuhkan Kesadaran Ecological Citizenship Pada Masyarakat Daerah Aliran Sungai Brantas (Studi Kass Kecamatan Wringinanom Kabupaten Gresik). *Kajian Moral dan Kewarganegaraan*, 7(2), pp. 1008–1002.

Development of road infrastructure to support the development of village tourism

H. Siswanto*, Pranoto, C.P. Dewi & B. Supriyanto
Universitas Negeri Malang, Malang, Indonesia

ABSTRACT: Selorejo Village of Dau subdistrict in Malang Regency is a developing tourism village. One of the problems faced by this village is the lack of infrastructure to support tourism development. The entrances to the tourism spots are mostly dirt and narrow roads. The purpose of this paper was to develop planning for road infrastructure to support the development of tourism in and around the village. The method used was field surveys such as land survey, soil carrying capacity survey, and traffic surveys. The planning included pavement thickness and pavement width. The type of pavement chosen was block pavement. Block pavements were chosen because it allowed accommodation of aesthetic elements and are easy to implement and maintain.

Keywords: infrastructure, tourism village, carrying capacity

1   INTRODUCTION

Road infrastructure development is a very strategic choice for supporting tourism sector development activities. Selorejo Village of Dau District in Malang Regency has many tourism potentials, such as citrus picking tours, Bedengan camping grounds, waterfalls, and others. Bedengan Tourist attractions are mostly camping grounds, and they are to be developed to attract tourist visits. The route to such sites requires adequate road infrastructure for safe and comfortable passage. The road to such locations is made up of uneven and narrow dirt roads, putting tourists at unnecessary safety risks. Apart from still being dirt roads, the passages to Bedengan sites are narrow, limiting its access to smaller vehicles. To develop tourism in Bedengan, infrastructure improvements should be prioritized.

According to the World Economic Forum (2019), infrastructure is part of the main sub-index that supports tourism businesses and natural and cultural resources, tourism policies, and supporting environments. The infrastructure sub-index includes air transport infrastructure, roads and port infrastructure, and tourism services infrastructures. Several researchers have conveyed the relationships between infrastructure and tourism. The availability of infrastructure is a determinant in attracting tourist visits (Boopen 2006; Khadaroo & Seetanah 2007; Brida et al. 2014; Alex-Onyeocha et al. 2015; Jovanović & Ilic 2016). Mandic et al. (2018) stated that tourist visit numbers are influenced by tourism infrastructure and facilities.

Infrastructure projects continue to be carried out by the Indonesian government to develop economic activities, including tourism. Tourism development has become part of the national priority programs (Moerwanto & Junoasmono 2017). The tourism potential of Indonesia is large, but tourist visits have not been maximized. One way to increase tourist visits is by increasing the availability of road infrastructure.

The development of road infrastructure will increase the index of Indonesia's tourism competitiveness, where Indonesia's index level is currently ranked 40th out of 140 countries (World Economic Forum 2019). In ASEAN countries, Indonesia's tourism competitive index in 2019 is

---

*Corresponding author: henri.siswanto.ft@um.ac.id

DOI 10.1201/9781003189206-13

inferior to Thailand's 31, Malaysia's 29, and Singapore's 17. Indonesia's rank has improved significantly compared to 4 years ago, where Indonesia sat at 50th place (Crotti & Misrahi 2017). Indonesia's position rose 10 levels from rank 50 to 40, which is the highest among ASEAN countries, followed by Thailand, which rose by 4 levels from rank 35 (Crotti & Misrahi 2017) to 31, while Singapore fell 6 ranks from rank 11 to rank 17, and Malaysia fell 4 levels from rank 25 to rank 29. The rank increase in the competitive tourism index results from the continuous and comprehensive efforts of the Indonesian Government in developing the infrastructure sector. The development of road infrastructure was carried out thoroughly and included village areas that could attract higher tourist visits. Based on these conditions, it was necessary to plan adequate, comfortable, and safe road infrastructure for tourists.

2 METHODS

2.1 *Preliminary surveys*

In preliminary surveys, field visits and coordination with the head of the village of Selorejo were conducted. Next, digging information related to the existing tourism potential of the village was conducted. Then, learning the potential and the existing issue to use them as the research materials and the solution-finding. The team that conducts preliminary surveys was the whole community service implementation team.

2.2 *Field data collecting*

The Primary data collecting includes soil bearing capacity survey, traffic survey, and land measurement survey. Soil bearing capacity data is obtained by conducting DCPT (dutch cone penetration test) in the field. This tool is available in the road pavement laboratory of the Department of Civil Engineering UM. An analysis was done to determine the points of the area to be tested. The test is done on several points between the village office to Bedengan tourist location. The number of points that are chosen is 5 points, proportional to the field analysis of the visual conditions of the road. To conduct this soil bearing capacity survey, DPCT form is used. This DPCT survey needs four DCPT operators in addition to the community service team.

The traffic data were obtained by conducting a field survey on roads heading to Bedengan tourist location. Survey points are first determined to be able to retrieve data that is representative and to be able to conduct the survey technically and easily. The points that are chosen are one point at the village entrance and one point in the road heading to the orange picking tourist location. To conduct this survey, two people need to remain on the spot at each point during the survey. The survey is held from 6.00 in the morning to 17.00 in the afternoon. The land measurement survey was conducted to obtain cross-sectional data of the road of the community service location. The survey is held in two days. The operators come from the Survey and Mapping Laboratory of the Department of Civil Engineering, and the survey helpers are four university students.

2.3 *Data processing and data analysis*

Data processing was conducted on soil bearing capacity data, traffic data, and land measurement data. The soil bearing capacity data contained in the survey forms are processed by inserting them into tables and calculations to obtain the value of the soil bearing capacity in the form of the CBR (California Bearing Ratio) value. Meanwhile, the survey forms' traffic data were processed to obtain the average daily traffic (ADT).

2.4 *Pavement design*

Based on the survey results' primary data, pavement design for this tourist village road can be arranged. There are three possible types of pavement: flexible pavement, rigid pavement, and concrete block pavement (CBP). The last type of pavement, which is the concrete block pavement, is chosen for this planning. This is based on the consideration of what use this road will have, which is for tourism. Therefore, the element of beauty must be prioritized, in addition to the elements of strength, workability, cost, and ease of maintenance. The method used in this pavement design was the Indonesian method or the component analysis method. The CBP design concept is the same as the pavement asphalt design (Hein & Smith 2014).

3 RESULTS AND DISCUSSIONS

In the preliminary survey, field visits were carried out to coordinate with village officials and determine the location for collecting technical data. The road conditions to the tourist location can be seen in Figure 1. The technical data that has been collected in this activity are soil bearing capacity data, traffic data, and land measurement data. Daily traffic data is presented in Table 1, while data on the soil bearing capacity is presented in Table 2. The soil bearing capacity for this community service location is soil with moderate bearing capacity, specifically with an average CBR (California Bearing Ratio) value of 7.28%.

The CBR founded allows all types of pavement to be applied. Three types of pavement can be applied, namely, flexible pavement, rigid pavement, and CBP. All types of pavement have their respective advantages and disadvantages. According to Gogoi (2019), choosing the type of pavement can be done based on, among others, the purpose of road services, construction implementation, maintenance, and availability of pavement material.

CBP has the advantage that it is easier to design, easy to procure, and easy to construct. These characteristics are suitable for road infrastructure in tourist areas. The usage of CBP is prevalent. According to Hettiarachchi and Mampearachchi (2016), CBP is in great demand due to its high availability in the market and its attractive design variations. In the last 10 years, CBP has been widely used in Europe because it is more durable against extreme weather. Also, it is more versatile compared to conventional asphalt pavement (Gunatilae & Mampearachchi 2019). CBP is also used for local road, urban roads, (Di Mascio et al. 2019), pedestrian path, bike lane, the road in the housing area, and parking lots (Miccoli et al. 2014). In fact, CBP is also used in airport runways and ports (Pradena & Houben 2016).

Table 1. The number of average daily traffic.

Vehicle Type	Volume (Vehicle per day)
Motorcycle	2397
Passenger car	489
Pickup truck	62
Bus	0
Truck	1

Table 2. Data showing capacity of soil.

Location test	California Bearing Ratio (%)
1	4.23
2	6.96
3	10.85
4	6.94
5	6.91

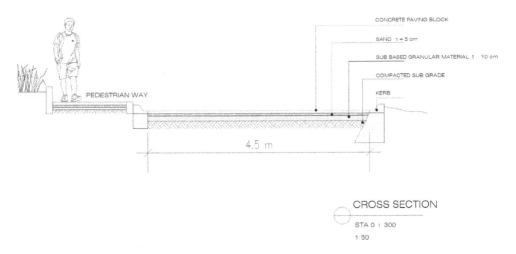

CONCRETE PAVING BLOCK

SAND : t = 5 cm

SUB BASED GRANULAR MATERIAL t 10 cm

COMPACTED SUB GRADE

PEDESTRIAN WAY

KERB

4.5 m

CROSS SECTION

STA 0 + 300

1:50

Figure 1.   Cross section.

The results of this road infrastructure design are presented in Figure 1. The width of the existing road is 2.2 m, which then widened twofold to 4.5 m. The purpose of the widening is to accommodate traffic that is originally one-way into two-way. A further widening of more than the mentioned width is not possible due to its high construction cost, especially the costs of soil excavation and backfilling work. A pedestrian walkway is also added. The pedestrians are provided with a separate walkway to make the tourist village visitors feel more comfortable and safer in walking.

In line with the purpose of the road service, which is for tourism, the element of beauty must be prioritized and the elements of strength, workability, cost, and ease of maintenance.

4   CONCLUSIONS

The type of pavement chosen in this road infrastructure design is the concrete block pavement. This is based on the purpose of the road, which is for tourism. Therefore, the element of beauty must be prioritized, in addition to the elements of strength, workability, cost, and ease of maintenance.

REFERENCES

Alex-Onyeocha, O. 2015. The impact of road transportation infrastructure on tourism development in Nigeria, *Pearl Journal of Management, Social Science and Humanities*, 1(2), pp. 48–55.
Boopen, S. 2006. Transport capital as a determinant of tourism development: A time series approach, *Tourismos*, 1(1), pp. 55–73.
Brida, J. G., Deidda, M. and Pulina, M. 2014. Tourism and transport systems in mountain environments: Analysis of the economic efficiency of cableways in South Tyrol, *Journal of Transport Geography*. Elsevier Ltd, 36, pp. 1–11. doi: 10.1016/j.jtrangeo.2014.02.004.
Crotti, R. and Misrahi, T. 2017. *Tourism Competitiveness - Paving the way for a more sustainable and inclusive future*. Available at: http://www3.weforum.org/docs/WEF_TTCR_2017_web_0401.pdf.
Gogoi, R. 2019. Cost effectiveness of interlocking concrete block pavements for low volume traffic roads, *International Journal of Engineering and Advanced Technology*, 8(6), pp. 1239–1244. doi: 10.35940/ijeat.F8382.088619.
Gunatilake, D. and Mampearachchi, W. K. 2019. Finite element modelling approach to determine optimum dimensions for interlocking concrete blocks used for road paving, *Road Materials and Pavement Design*. Taylor & Francis, 20(2), pp. 280–296. doi: 10.1080/14680629.2017.1385512.

Hein, D. K. and Smith, D. R. 2014. Development of an asce standard for permeable interlocking concrete pavement, *2014 Transportation Association of Canada Conference and Exhibition: Past, Present, Future, ATC 2014*, (October 2015).

Hettiarachchi, H. A. C. K. and Mampearachchi, W. K. 2016. New block design and laying parameters for interlocking concrete block pavements to improve human thermal comfort levels in urban spaces, *International Journal of Sustainable Building Technology and Urban Development*, 7(2), pp. 104–115. doi: 10.1080/2093761X.2016.1172278.

Jovanović, S. and Ilic, I. 2016. Infrastructure as important Determinant of Tourism Development in The Countries of Southeast Europe, 5(1), pp. 288–294.

Khadaroo, J. and Seetanah, B. 2007. Transport infrastructure and tourism development, *Annals of Tourism Research*, 34(4), pp. 1021–1032. doi: 10.1016/j.annals.2007.05.010.

Mandic, A., Mrnjavac, Ž. and Kordic, L. 2018. Tourism infrastructure, recreational facilities and tourism development, *Tourism and Hospitality Management*, 24(1), pp. 41–62. doi: 10.20867/thm.24.1.12.

Di Mascio, P., Moretti, L. and Capannolo, A. 2019. Concrete block pavements in urban and local roads: Analysis of stress-strain condition and proposal for a catalogue, *Journal of Traffic and Transportation Engineering (English Edition)*. Elsevier Ltd, 6(6), pp. 557–566. doi: 10.1016/j.jtte.2018.06.003.

Miccoli, S., Finucci, F. and Murro, R. 2014. Criteria and procedures for regional environmental regeneration: A European strategic project, *Applied Mechanics and Materials*, 675–677(December), pp. 401–405. doi: 10.4028/www.scientific.net/AMM.675-677.401.

Moerwanto, A. S. and Junoasmono, T. 2017. Integrated Turism Insfrastructure development Strategic, *HPJI Jounal*, 3(2), pp. 67–78. doi: 10.26593/.v3i2.2735.%p.

Pradena, M. and Houben, L. 2016. Functional criteria for sustainable design of urban pavements, *Gradjevinar*, 68(6), pp. 485–492. doi: 10.14256/JCE.1464.2015.

World Economic Forum. 2019. *The Travel and Tourism Competitiveness Report 2019*.

Community Empowerment through Research, Innovation and Open Access – Sayono et al (Eds)
© 2021 Copyright the Author(s), ISBN 978-1-032-03819-3

Problems of online learning in a borderland of Indonesia in the COVID-19 pandemic

A. Sultoni*, E.P. Priyatni, A.M. Nasih, A. Zahro, Juharyanto & Ibrohim
Universitas Negeri Malang, Malang, Indonesia

ABSTRACT: The COVID-19 pandemic in Indonesia has altered dramatically many aspects of life. In formal education, the Ministry of Education issued a policy instructing schooling sectors to implement online learning to prevent students from contracting COVID-19. Therefore, the aim of this study was to portray problems emerging in online learning at senior high schools in a borderland of Indonesia: the province of North Kalimantan. Data was collected in September 2020 through an online open questionnaire asking questions to 251 teachers at senior high schools throughout the regencies. Data was analyzed through descriptive quantitative and qualitative frameworks. The results revealed that the emerging problems of online learning included internet networks, internet data, learning devices, student attitudes, parent attitudes, learning evaluations, teaching methods, the literacy of information technology, and student comprehension. To conclude, the most noticeable problems are internet access, competency in online learning, and student attitudes.

Keywords: online learning, borderland, COVID-19 pandemic, Senior High School

1 INTRODUCTION

Coronavirus disease (COVID-19) was first reported in Wuhan, China, in December 2019 by Wuhan Municipal Health Commission. This deadly virus transmits from person to person and has killed millions of people and became a pandemic in more than 190 countries (Li et al. 2020; Huang et al. 2020; WHO 2020). In Indonesia, the case of COVID-19 is first confirmed in March 2020 and then infected more than 433,000 people. The virus spread over all provinces and caused lockdown in many provinces (Gugus Tugas 2020).

The COVID-19 pandemic in Indonesia drastically changed many aspects of life, not only in the economy but also in tourism, religion, education and society. In education, the Ministry of Education and Culture closed almost all schools and implemented online learning to prevent COVID-19 transmission (Kemdikbud 2020; media Indonesia 2020). The implementation of online learning at a national scale brought opportunity (Gonzalez et al. 2020) and hindrance (Yang 2020). However, the implementation of emergency online learning in remote areas seemed to cause more hindrance and problems than opportunities.

Online learning is a kind of learning which depends on the availability of information technology. Internet access and technological devices, such as smartphones and computers are crucial because without them, it is likely impossible to implement online learning (Singh & Thurman 2019; Adedoyin & Soykan 2020). Digital competence in operating the information technology plays important roles in effective online learning (Ferrari 2012). Another aspect that affected online learning is teacher competency in designing effective and interesting online learning. One of the crucial issues is how to control cheating with the limited supervision of teachers (Arkorful & Abaidoo 2014).

---

*Corresponding author: achmad.sultoni.fs@um.ac.id

DOI 10.1201/9781003189206-14   69

Activities of teaching and learning in online learning can be divided into synchronous and asynchronous. Synchronous learning using video conference tools (such as Zoom and Google Meet) needs good internet connectivity and comes at a high cost. However, asynchronous learning through videos of recorded lectures or PowerPoint presentations costs less money and does not always require good internet access (Azlan et al. 2020; Crawford et al. 2020; Patricia 2020).

Problems and hindrances in online learning due to the COVID-19 pandemic must be urgently investigated. The study's focus is not only limited to kind of problems, but also the intensity of problems and objects of problems. The investigation results are important for the government to provide policies and rules of effective education for citizens. Studies on problems of online learning in Indonesia are abundant, some focused on university (Hamid et al. 2020; Rachmawati et al. 2020, Agung et al. 2020), in elementary schools (Purwanto et al. 2020, Putria et al. 2020), and for disabled students (Rof'ah et al. 2020). Studies on the problem of online learning in remote areas seem sparse both in number and scope. For instance, online learning problems at elementary schools in a rural area of Bojonegoro, East Java (Setiawan & Iasha 2020), and comparisons of online learning in urban and remote areas are limited (Sulisworo et al. 2020).

The present study investigated online learning problems in senior high school due to the COVID-19 pandemic in Indonesia's borderland, namely North Kalimantan. This location was chosen to portray the worst effects of emergency online learning in Indonesia since North Kalimantan is the newest province in Indonesia (established in 2013) and is located in a relatively remote area near Malaysia (Kaltaraprov 2020).

2 METHODS

In this study, the subjects were 251 teachers at senior high school in North Kalimantan. All have obtained the certificate of a professional educator. They are largely teachers at public schools, both vocational and non-vocational schools, throughout North Kalimantan (Tarakan, Bulungan, Malinau, Nunukan, and Tana Tidung). Their last education mainly included graduate level courses though some graduated from postgraduate studies and have taught for more than ten years on average.

Data was collected in September 2020 through open questionnaire via a Google form. The respondents could answer more than one problem of online learning they encountered. All data was analyzed through qualitative analysis (Miles et al. 2014) and simple a descriptive statistic. At first, all data was classified according to types of online learning problems. Then the percentage of each problem is formulated and a conclusion is drawn.

3 RESULTS AND DISCUSSION

The respondents wrote 485 problems emerging in online learning, which are classified into nine problems. The problems consisted of internet access and devices, student and parent attitudes, and literacy in online learning. Table 1 explains the problems.

Table 2 shows that the problems of online learning due to the COVID-19 pandemic consist of the availability of the internet network, the unavailability of internet data, the unavailability of learning devices, student attitudes, parent attitudes, the evaluation of learning, the teaching method, the literacy of technology and student understanding. It is obvious from Table 2 that the main problems of online learning in the borderland are internet networks, internet data and student attitudes. On the contrary, parent attitudes, technological literacy and student understanding are minor problems with minimal occurrences. In detail, the percentage of each online learning problem is described in figure 1.

All problems of online learning described in Figure 1 can be classified in a more concise way. Problems of internet network and internet data are related to internet access, whereas problems of

Table 1. The types of problems of online learning in the borderland.

No	Type of problems	Occurrence	Details
1	Internet network	138 times	Mainly unstable internet networks. In some cases, there is no network at all.
2	Internet data	91 times	Students did not have internet data or Wi-Fi access. In some towns, sometimes electricity is off.
3	Learning devices	58 times	Students did not have smartphones or laptops due to economic factors.
4	Attitude of students	102 times	This problem includes absences in a virtual class, lack of motivation, and lack of discipline in joining classes and doing homework. Students also did not do homework, got bored, or give few responses in a virtual class.
5	Attitude of parents	8 times	Some parents pay no attention to online learning, and even ask students to work with them.
6	Learning evaluation	22 times	Teachers find it is difficult to evaluate students' achievements, especially in the aspect of moral and skill.
7	Teaching method	52 times	It is hard for teachers to determine effective ways/methods in teaching and to manage time well. Teachers also cannot control students' learning activities. These conditions, in turn, affect the achievement of learning objectives.
8	Literacy of technology	6 times	Some teachers and students did not understand the technology of online learning.
9	Students understanding	8 times	Students cannot understand the teaching material.
		Total number of problems: 485	

Table 2. Percentage of online learning problems in the borderland.

Number	Type of Problems	Occurrence	Percentage
1	internet network	138	28.45%
2	internet data	91	18.76%
3	learning devices	58	11.96%
4	student attitude	102	21.03%
5	parent attitude	8	1.65%
6	learning evaluation	22	4.54%
7	teaching method	52	10.72%
8	literacy of technology	6	1.24%
9	students understanding	8	1.65%
	Total:	485	100%

learning evaluation, teaching methods, literacy of information technology, and student understanding can be categorized as the competency of online learning. In other words, problems of online learning consist of internet access, learning devices, the competency of online learning, student attitudes and parent attitudes. The changes in problem types, in turn, change the percentage of some problems. Internet access becomes the biggest problem of online learning in North Kalimantan. The changes are portrayed in Figure 1.

Nine problems emerging in emergency online learning in North Kalimantan are possibly related to each other. Student difficulties in accessing the internet (i.e., unstable internet network, no internet data, and no learning devices) negatively affect their attitude to online learning; for instance, they

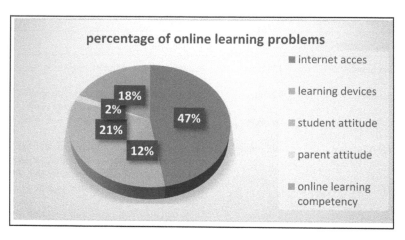

Figure 1.   Types of online problems and their percentage in brief.

were frequently absent in online learning and were acting indisciplined. This phrase corresponds to Dhawan's (2020) statement that students with no internet access and no smartphones/laptops are in big trouble in online learning, since they lose the opportunity for learning. Adedoyin & Soykan (2020) explained that internet access's main factors are socio-economic and relating to the internet network. Students from low socio-economic groups will be left behind in online learning.

That some teachers in North Kalimantan are less competent in online learning due to the COVID-19 pandemic confirms previous research results (Setiawan & Iasha 2020; König et al. 2020). König et al. reported that new teachers (that are "digital natives") in Germany have not developed digital ability in online teaching. Moreover, many schools in Germany are left behind in implementing information and communication technology (ICT). The finding is also in accordance with Dhawan's (2020) statement that it is hard for teachers to change mode, methodology and time management from offline to online learning. In addition, Adedoyin & Soykan (2020) said that most teachers are not ready to design qualified learning activities.

Internet access becomes the main problem of online learning in remote areas such as in North Kalimantan. Previous research by Setiawan and Iasha (2020) showed that the big problem of online learning in a rural area of Bojonegoro was an unstable internet network. The condition also happens beyond COVID-19 pandemic (Rao et al. 2011). The statement corresponds with a previous study that online learning's common problem in remote areas is a slow internet connection (AISR 2006) and information technology mastery (Ramanujam 2002).

Other finding indicated that students perceived negatively online learning corroborates research in Ghana. Aboagye et al. (2020) found that students in Ghana would choose to cancel online learning in the COVID-19 pandemic if they could. The negative attitude to online learning is correlated with a middle to a high economic level. Sindiani et al. (2020) reported that medical students in Jordan preferred traditional face-to-face teaching to online teaching. Similarly, students at dental school in China had low progress in synchronous online prosthodontic courses (Yang 2020). Surrounding distractions may cause this negative attitude, such as: difficulty in accessing the internet, the virtual classroom, low motivation, low engagement to online learning, deficient smart phones/laptops and inadequate competency in using information technology (Elahi 2012; Zhang et al. 2020).On the contrary, research by Qazi et al. (2020) revealed that students in Brunei and Pakistan, with ease of access to the internet and affordability of mobile phones considered online learning to be a positive thing. This positive attitude is partly because of students' intention and good acceptance (Yakubu & Dasuki 2019; Kemp et al. 2019; Tarhini et al. 2016).

4 CONCLUSION

The emerging problems of sudden online learning in borderland (North Kalimantan) are internet access (internet network and internet data), competency in online learning, student attitudes, learning devices, and parent attitudes. The most noticeable problems are the access of the internet and competency of online learning. As an implication, the government is encouraged to build an infrastructure of the internet connection in North Kalimantan and hold training in online learning to provide qualified education in remote regions.

REFERENCES

Aboagye, Emmanuel, Joseph Anthony Yawson, Kofi Nyantakyi Appiah. 2020. COVID-19 and E-Learning: The Challenges of Students in Tertiary Institutions. *Social Educa-tion Research*, Volume 1 Issue 1, 109-115, DOI: https://doi.org/10.37256/ser.122020422.

Adedoyin, Olasile Babatunde, & Emrah Soykan. 2020. COVID-19 pandemic and online learning: the challenges and opportunities, Interactive Learning Environments, DOI: 10.1080/10494820.2020.1813180

Agung, Antonius Setyawan Sugeng Nur, Monika Widyastuti Surtikanti, & Charito A Quinones. 2020. Students' Perception of Online Learning during COVID-19 Pandemic: A Case Study on the English Students of STKIP Pamane Talino. *SOSHUM Jurnal Sosial dan Humaniora [Journal of Social Sciences and Humanities]* Volume 10, Number 2, 2020.

Arkorful, V., & Abaidoo, N. 2014. The Role Of E-Learning: Advantages and Disadvantages of Its Adoption in Higher Education. International Journal of Education and Research, 2 (12), 397 – 410.

Australian Institute for Social Research (AISR). *The Digital Divide: Barriers to E-learning*. Final Report presented to Digital Bridge Unit, Science Technology and Innovation Directorate. University of Adelaide, Australia, 2006.

Azlan, Che Ahmad, Jeannie Hsiu Ding Wong, Li Kuo Tan,Muhammad Shahrun Nizam A.D. Huri,Ngie Min Ung, Vinod Pallath,Christina Phoay Lay Tan,Chai Hong Yeong, Kwan Hoong Ng. 2020. Teaching and learning of postgraduate medical physics using Internet-based e-learning during the COVID-19 pandemic – A case study from Malay-sia. *Physica Medica,* 80 (2020) 10–16, Published by Elsevier, https://doi.org/10.1016/j.ejmp.2020.10.002.

Crawford, J., Butler-Henderson, K., Rudolph, J., & Glowatz, M. 2020. COVID-19: 20 Countries' Higher Education Intra-Period Digital Pedagogy Responses. *Journal of Applied Teaching and Learning (JALT)*, 3(1). https://doi.org/10.37074/jalt.2020.3.1.7

Dhawan, Shivangi. 2020. Online Learning: A Panacea in the Time of COVID-19 Crisis. *Journal of Educational Technology Systems,* Vol. 49(1) 5–22, DOI: 10.1177/0047239520934018.

Elahi, Rashid M Uzma. 2012. Use of educational technology in promoting distance education. *Turkish Online Journal of Distance Education-TOJDE*, January, Volume: 13 Number: 1 Article 3; :79-86. ISSN 1302-6488.

Ferrari, Anusca. 2012. *Digital Competence in Practice: An Analysis of Frameworks*. Euro-pean Commission Joint Research Centre, Institute for Prospective Technological Studies, doi:10.2791/82116.

Gonzalez T, de la Rubia MA, Hincz KP, Comas-Lopez M, Subirats L, Fort S, et al. 2020. Influence of COVID-19 confinement on students' performance in higher education. *PLoS ONE* 15(10): e0239490. https://doi.org/10.1371/journal.pone.0239490

Huang, C., Wang, Y., Li, X., Ren, L., Zhao, J., Hu, Y., Zhang, L., Fan, G., Xu, J., Gu, X., & Cheng, Z. 2020. Clinical features of patients infected with 2019 novel coronavirus in Wuhan, China. *The Lancet,* 395(10223), 497–506. https://doi.org/10.1016/S0140-6736(20)30183-5

Kemp, A., Palmer, E., and Strelan, P. (2019), A taxonomy of factors affecting attitudes towards educational technologies for use with technology acceptance models. *British Journal Education Technology*, 50, 2394–2413. doi:10.1111/bjet.12833

König, Johannes, Daniela J. Jäger-Biela & Nina Glutsch. 2020. Adapting to Online Teaching During COVID-19 School Closure: Teacher Education and Teacher Competence Effects Among Early Career Teachers in Germany. *European Journal of Teacher Education*, 43:4, 608–622, DOI: 10.1080/02619768.2020.1809650.

Li, Q., Guan, X., Wu, P., Wang, X., Zhou, L., Tong, Y., Ren, R., Leung, K. S., Lau, E. H., Wong, J. Y., & Xing, X. (2020). Early transmission dynamics in Wuhan, China, of novel coronavirus–infected pneumonia. New England *Journal of Medicine,* 382,1199–1207. https://doi.org/10.1056/NEJMoa2001316

Miles, M. B., Huberman, A. M. & Saldana, J. 2014. *Qualitative data analysis: An expanded sourcebook* (3rd ed.). London: SAGE.

Patricia. A. 2020. College Student' Use and Acceptance of Emergency Online Learning Due to COVID-19, *International Journal of Educational Research Open*, doi: https://doi.org/10.1016/j.ijedro.2020.10001

Putria, Hilna, & Luthfi Hamdani Maula, Din Azwar Uswatun. 2020. Analisis Proses Pembelajaran dalam Jaringan (DARING) Masa Pandemi COVID-19 Pada Guru Sekolah Dasar. *Jurnal Basicedu,* Vol 4 No 4 2020, 861–872. DOI: https://doi.org/10.31004/basicedu.v4i4.460.

Ramanujam, P. Renga. 2002. *Distance Open Learning: Challenges to Developing Countries.* Delhi, India: Shipra Publications.

Rao, K., Eady, M., & Edelen-Smith, P. 2011. Creating virtual classrooms for rural and remote communities. *Phi Delta Kappan.* 92(6), 22–27.

Setiawan, Bramianto & Vina Iasha. 2020. COVID-19 Pandemic: The Influence of Full-Online Learning for Elementary School in Rural Areas. *Jurnal Pendidikan Sekolah Dasar (JPSD)*, Vol. 6 No. 2, September 2020, 114–123. DOI: Http://Dx.Doi.Org/10.30870/Jpsd.V6i2.8400.

Sindiani, Amer Mahmoud, Nail Obeidat, Eman Alshdaifat, Lina Elsalem, Mustafa M.Alwani, Hasan Rawashdeh, Ahmad S.Fares, Tamara Alalawne, and Loai Issa Tawalbeh. 2020. Distance education during the COVID-19 outbreak: A cross-sectional study among medical students in North of Jordan. *Annals of Medicine and Surgery* 59 (2020) 186–194, https://doi.org/10.1016/j.amsu.2020.09.036.

Sulisworo, Dwi, Dian Artha Kusumaningtyas, Agnes Bergita Anomeisa, Wahyuningsih & Widya Rahmadhani. Perceptions of Online Learning Experiences Between Students in Urban and Remote Areas: Case Study in Indonesia. *International Journal of Scientific & Technology Research*, vol 9, Issue 02, February 2020, 4850-4854, Issn 2277-8616 4850 IJSTR©2020.

Tarhini, A., Hone, K., Liu, X., & Tarhini, T. 2016. Examining the moderating effect of individual-level cultural values on users' acceptance of E-learning in developing countries: a structural equation modeling of an extended technology acceptance model. *Interactive Learning Environments*, 25(3), 306-328, doi:10.1080/10494820.2015.1122635

Vandana Singh & Alexander Thurman. 2019. How Many Ways Can We Define Online Learning? A Systematic Literature Review of Definitions of Online Learning (1988-2018), *American Journal of Distance Education*, 33:4, 289-306, DOI: 10.1080/08923647.2019.1663082

Yakubu, M. N., & Dasuki, S. I. 2019. Factors affecting the adoption of e-learning technologies among higher education students in Nigeria: A structural equation modeling approach. *Information Development*, 35(3), 492–502. https://doi.org/10.1177/0266666918765907

Yang, Xu, Deli Li, Xiaoqiang Liu, Jianguo Tan. 2020. Learner behaviors in synchronous online prosthodontic education during the 2020 COVID-19 pandemic, *The Journal of Prosthetic Dentistry,* ISSN 0022-3913, https://doi.org/10.1016/j.prosdent.2020.08.004.

Zhang, W.; Wang, Y.; Yang, L.; Wang, C. 2020. Suspending Classes Without Stopping Learning: China's Education Emergency Management Policy in the COVID-19 Outbreak. *J. Risk Financial Manag. 13*, 55.

https://covid19.go.id/peta-sebaran

https://kaltaraprov.go.id/profil/sejarah.

https://mediaindonesia.com/read/detail/327729-tegakkan-ketentuan-sesuai-skb-4-menteri

https://www.kemdikbud.go.id/main/blog/2020/08/penyesuaian-keputusan-bersama-empat-menteri-tentang-panduan-pembelajaran-di-masa-pandemi-covid19

https://www.who.int/emergencies/diseases/novel-coronavirus-2019

Community Empowerment through Research, Innovation and Open Access – Sayono et al (Eds)
© 2021 Copyright the Author(s), ISBN 978-1-032-03819-3

Integrating gender and religion: Social transformation for strengthening identity among Indonesian Muslim Migrant Workers

Anggaunitakiranantika* & P.P. Anzari
Universitas Negeri Malang, Malang, Indonesia

ABSTRACT: In Indonesia, as the highest Muslim population country worldwide, patriarchy systems are correlated with Islamic norms and values used for controlling human behavior, including the relationships of men and women. Addressing this issues, Muslim women try to expand their capabilities by migrating to Hong Kong as migrant workers. This research aims to study Indonesian Muslim women in Hong Kong based on their gender identity and religious value. A qualitative method was used. Applying *purposive samples*, research focused on Indonesian migrant workers in Hong Kong who joined a Muslim community. Research found that women's transformation is strengthened by gender and religion, which used for reinforcing as Muslim Indonesian migrant worker activities and Muslim women community practices among Javanese Muslim women at Hong Kong.

Research found that migration roles as a part of Muslim women new value, which is give a flexibility for Muslim women for working abroad as the transformation. In addition, Muslim women transformation in Hong Kong not only interpreted as individual norms but also manifested in collective activities during their leisure time in Hong Kong for strengthening Muslims women identity.

Keywords: gender, religion, social transformation, Indonesian migrant worker

1 INTRODUCTION

International migration by Indonesian women is increasingly prevalent and is carried out to diverse Asian countries. Increased demands and living costs trigger the increase of Indonesian women entering the public sector by becoming Indonesian Migrant Workers (IMW). Ionesco and Agzaharn (2009) stated that male and female relations are affected by living expense changes in society. As a result, people are forced to migrate according to gendered and geographical factors, which are governed by sets of laws and regulations from each country. Furthermore, migration often implies economic and social costs and benefits for both males and females, who participate in international migration.

East Java is the one of the provinces in Indonesia that contributes the highest number of migrant workers in several Asian countries. East Java provides Muslim laborers that migrate to many Asian countries such as Hong Kong, Singapore, Malaysia, Taiwan, and South Korea. A small part of the data on Indonesian labor shows increasing interest in earning from other countries, despite the risks to migrant workers. One of the attractive reasons for short term international migration is the success of fellow migrants measured by the possession of financial remuneration or remittance that they send to their home regions (Constable, 2010; Anggaunitakiranantika, 2016; Hamidi, 2017; Allmark & Wahyudi, 2019).

Intersectionality factors against Javanese women who work as migrant workers in Hong Kong explain the diverse socially and culturally constructed categories which interact at different levels

---

*Corresponding author: anggaunita.fis@um.ac.id

to produce different forms of relations and inequalities (Kiranantika, 2020). The social and cultural constructions which are recognized as patriarchy systems tend to put women in an unequal position like patron-client, labor worker-employer, labor worker-recruiting agency, employer-recruiting agency, and local worker-migrant worker (from different regions). The inequality commonly faced by Indonesia's female migrant workers are exploitation, remuneration which is below the standard of ILO's labor regulations, abuse (physical, verbal, and psychological) against migrant workers, negative stereotypes and stigma regarding Muslim migrant workers, and Islamophobia triggered by any Muslim-related activities conducted by migrant workers in Hong Kong.

Meanwhile, the preliminary studies showed that Muslim women and Indonesian migrant workers are still bound to their responsibilities of doing worship appropriately, despite being away from their country and living in different surroundings in Hong Kong. Interestingly, by being religious minorities in Hong Kong, they are involved in preaching and Islamic Hadith interpretation as a weekly agenda for strengthening their identity among Hong Kong peoples. By elaborating on gender perspectives for pointing out that women have their own life meanings and other capabilities instead of just being workers, the Muslim communities of Indonesian women workers in Hong Kong tend to build their identity through positive attitudes. Commonly, researchers limit their scope of observations on international migrant workers on remittance and migration experience by socioeconomic conditions. This paper will be analyzing how Indonesian Muslim women who work as migrant workers perceive the gendered notion regarding their migration to Hong Kong and identifying the effects of Muslim spiritual activities in Hong Kong as new residents which different from the Chinese traditions there.

2 METHODS

This study employed a qualitative method with a phenomenology as the design for data gathering. It was used to captured written and oral descriptive data from people and observable behavior in Hong Kong. This method was chosen because the present study examines holistic, complex, dynamic, and meaningful problems. In addition, the qualitative method is part of the knowledge that can be considered a social product and a social process (Neuman, 2000). The data was collected from female Indonesian Muslim migrant workers working in Hong Kong. Several steps were used in collecting the data, namely: reading critically, making notes, conducting participant observation, analyzing the data, and concluding the substance of the study (Miles & Huberman, 1994). Additionally, participant-observation amid Indonesian migrant worker organizations in Hong Kong was used for gathering information and collecting the additional data.

In addition, Sheller (2011) suggested that research in mobility allows to see what defines social science along with the material and political relations that construct the structure of the world. Therefore, the purpose of this study is to reflect the relationship between the observed and the observer. This study also seeks to uncover natural settings on weekly religious activities through social phenomena enacted by Indonesian Muslim women as Indonesian migrant workers in Hong Kong. Participants were recruited using a purposive sampling technique. The Javanese Muslim communities in Hong Kong are the biggest association for Muslim migrant workers from Indonesia. Further in-depth interview and observation were conducted to eleven Javanese Muslim women from Muslim communities in Hong Kong. They were all domestic workers and caregivers in Hong Kong and have worked for at least four years. This study was conducted from February to November 2019.

The research location was a Hong Kong district with many migrant workers (a melting pot), namely Causeway Bay district, that has been part of the public sphere in Hong Kong since early 2000. Causeway Bay has always been a chosen place for gathering, especially for women migrant workers from Indonesia. To obtain more information, two Muslim community leaders and an Indonesian migrant-employer in Hong Kong are chosen as research subjects for the phenomenon related to gender notions and religion activities, which are embedded in Indonesia migrant workers.

3 RESULTS AND DISCUSSION

3.1 *Gender belief systems among Indonesian Muslim women in Hong Kong*

Being migrant workers in Hong Kong is an initial step action for Indonesian Muslim women to reject discrimination and subordination which come from masculine superiority in Javanese societies or the so-called patriarchal system. Indonesian people are engaged in a patriarchal system which puts men as superior to women. It has been constructed for a long time and rigidly bound together men and women throughout time. Nowadays, this system has been changed by mobility. While mobility is considered a common thing for Indonesian men only, Indonesian women are considering to be living beyond the norms. Indonesian people, especially Javanese women, used to be allowed to nurture their family and work in domestic sectors only as migration might disrupt predominant Javanese cultural values.

Both Koning et al (2000) and Mahler and Pessar (2001) claimed that the discourse of womanhood in Indonesia, in this case named *ibuism*, and the discourse of traditional cultural values are ingrained in kin-based institutions in Indonesia, especially for Javanese people. As a result, also mentioned by William and Widodo (2009), the so-called 'feminine jobs' restrict the choices of occupation available for women. These conditions are still existing to this day. As in Javanese culture, women are still restricted from certain occupations despite their increased access to education as well as vocational and professional training, the two major forces in women lives. Javanese people believe that women must live more in the domestic life rather in the public.

Reflecting to this case, the process of Indonesian Muslim women embarking towards independence roots from living in a patriarchal system, wherein a father or husband uses political hegemony in treating women in the name of respecting the culture. Women are only permitted to perform activities in a restrictive sector, namely the domestic sector. This condition has aggravated the lower bargain position for women in Indonesia as this system sets women up to be taken for granted in family decision making. All decisions must be decided by a man. Furthermore, this custom is supported by the local bureaucrats in Indonesia who gain opportunities from this situation. To illustrate this case, when a woman wishes to work abroad, she must acquire the permissions from their spouse or father or uncle or brother. If this requirement is not included on the form to be submitted to the government, it will result in her paying more than women who include permissions from their male relatives. In other words, the requirements for engaging Indonesian women in work are very rigid and requires various recommendations from the spouse or father.

Based on the observation, women workers in this study are generally married and leave their children with their husbands and are supported also by their extended family. Javanese women usually sign four-to-eight-year contracts with their recruitment agency, a consideration amount due to family needs and basic needs. In addition, Javanese women who decided to be Indonesian migrant workers in Hong Kong also said that they really want to prove that they are capable of changing their future and, therefore, contradicting the patriarchal believes which are embedded into Javanese society, such as the idea that being a female migrant worker is associated with having low skills, being a needy and low educated person, and that they should stay at home for their family. By working abroad as migrant workers, Javanese women have clear goals in their mind for showing their capabilities, the ability to live independently by being away from their families and having working competence instead of adhering to the labels and stigmas from others in Indonesia given to them.

The result of this study is connected with a new direction in the theory of mobility and is also a response to several important feminist critiques of mobility theory that point out it is grounded in masculine subjectivity, makes assumptions about freedom of movement, and ignores the gendered production of space. Skeggs (2004) argued that the (old) mobility paradigm could be linked to a 'bourgeois masculine subjectivity' that describes itself as 'cosmopolitan' and pointed out that 'mobility and fixity are figured differently depending on national spaces and historical periods' (Skeggs, B. 2004). This is appropriate to the migration experience of Javanese women, particularly those who considered their mobility as an initiative of the patriarchal system as settled social structures. International migration performed by Javanese women acts as a form of liberation for women and a response to patriarchal culture which clings onto the subjectivity principle of supporting male dominance in social structure.

Ridgeway and Correll (2004) stated that one of the core elements which makes up the gender system is cultural beliefs and their impact, namely "social relational" impact. In Addition, when gender has a salient role in different contexts, cultural beliefs regarding gender systematically create bias on the behaviors, performance, and evaluations of men and women. In this research, Javanese women who work as migrant workers in Hong Kong showed rationality toward their migration decisions. They were only given the domestic sector to explore in Indonesia and once they migrated to Hong Kong, they got pay and the freedom to live independently. By migrating internationally, these Javanese women left their house to go to a faraway country for a long period of time and thus obtained salvation from being liberated from the culture that caused "scars", because the gender identity of Javanese women in Hong Kong was constructed by independence, knowledge, and insights from being migrant workers. Such rationality possessed by Javanese women makes the difference in the mobility dynamic context becoming more developed. By being Indonesian migrant workers in Hong Kong, Javanese women can obtain working experience and have broad knowledge in being true women in another country.

3.2 *Perpetuating minorities: Indonesian Muslim women's' journey in Hong Kong*

The period of social change is described by social anthropologists as a period of "liminality" in communities in which bonds are egalitarian and all possibilities are open. Turner (1974) argued that community is considered sacred or "holy" almost everywhere, possibly because it goes beyond the norms that rule structured and institutionalized relationships and is accompanied by experiences of unprecedented potency. Furthermore, Runnals (2002) stated that it is essential to study women's positions in religions regarding social systems which are significantly imbued with gender concepts that are related to divinity, considering that gender configurations are both behavioral and symbolic. The hegemonic system which delivered Javanese people in patriarchal system are also described in many ways and in many concepts.

On the other hand, Appadurai (2006) noted that "predatory identities," which stem from the small gap between majority status and total national ethnic purity, can be the source of extreme ire against certain ethnic others. The grounds for such incidents are the inner interactions between the majority and minority. Then, he further stated that, incensed by the fear of being driven to become minorities, and the fear that the minorities can readily take their place, majorities can be mobilized to react to perceived dangers (Appadurai, 2006). In line with this study, the holy activities of Indonesian Muslim women are seen when they have opportunities and freedom to express their activities. Notably, Hong Kong's culture and regulation allowed them to pay attention to some restrictions, but Indonesian Muslim women see this not as a barrier, but as a challenge of Muslim ritual activities abroad (Table 1).

For Muslim women in Hong Kong, specifically those who work as an Indonesian migrant worker, assembling together on the weekend was a form of a refreshment activity, strengthening their sense of nationalism as they live a Hong Kong multicultural life and is also a form of gratitude to God for the chance to live another day as a better person. Reflecting on the concept of reciprocity by Rapoport et al., (2002), it is unlikely that these female migrant workers can spend much time away from their work. Javanese migrant workers often work an unclear number of hours and do not have fixed job descriptions due to their being domestic workers in Hong Kong. Nevertheless, they still allocate time for conducting religious activities together with other Muslim women from Indonesia, despite the possibilities of them being called to come back to serving their employers. These religious activities are binding for migrant workers in whole Hong Kong districts. Not only for Javanese Muslim migrant workers around Causeway Bay district and Mong Kok district, but also in the Kowloon mosque which is the largest mosque in Hong Kong, and other mosques in other Hong Kong districts. Having such activities is meant for Muslim women as ways to feel the security of being minorities.

Table 1. Muslim Women Religious Activities in Hong Kong Table.

	Types of Activities	Time (Flexibility)	Place
Activity 1	Doing Recitation/Preaching	Weekly	Mosque, Hong Kong District public sphere area
Activity 2	*Tausiyah*/Islamic *Hadith* reviewing	Weekly/ Twice a month	Mosque, Hong Kong District public sphere area
Activity 3	*Ziarah* (pilgrimage) Tour to Islam community in Mainland China (Uyghur)	Quarter in a year	Certain place in Hong Kong District (Bus Station)
Activity 4	Singing Islamic songs, *Qasidah* and *Hadrah* (Singing Islamic songs accompanied by *Rabanna* and polite-slow motion)	Weekly/Twice a month	Hong Kong District public sphere area
Activity 5	Fund raising for *Shodaqoh* and *Infaq*	Weekly	Mosque, Hong Kong District public sphere area
Activity 6	*Buka Bersama, Shalah Taraweeh* (Fast-breaking together during Ramadhan followed by *shalah sunnah*)	Twice in Ramadhan Month	Certain place in Hong Kong District
Activity 7	Eid Celebration	Twice a year (Eid Fitri and Eid Adha)	Mosque, Hong Kong District public sphere area

In line with these research findings, Bertrand (2004); Hamidi (2017) and Anggaunitakiranantika & Hamidi (2020) explained that conflict might be triggered by certain economic drivers, but this can be exacerbated more when combined with differences in religious and ethnic identities. Conflicts have been faced by Javanese Muslim women with their families in Indonesia, but migration has offered better opportunities in working and public participation, including delivering the ideas to others and achieving women's goals. Research found that religious activities were mostly held during their leisure time in Hong Kong. It has been running this way for many years without any race differentiation and is open for any migrant worker in Hong Kong. Based on the activities among Javanese Muslim women in Hong Kong, only Muslim women from Indonesia could attending all types of religious activities in Hong Kong and they were only recognized as Indonesian Muslim women with their religious customs. In Hong Kong, Islamic identities are not only represented by their appearance, such as wearing hijab and long apparel, but also by their behaviors, such as eating and drinking halal foods and beverages. Integration sheds light into constructed categories which interact with, among others, gender beliefs and religion levels to produce different forms of power relations. Javanese Muslim integrity shows through such religious activities as the Indonesian Muslim migrant workers have the power to voice inequalities which arise due to conflicts and gaps among the patriarchal system in Indonesia and the participation of women in Hong Kong as migrant workers. Furthermore, the migration decision among Indonesian women somehow could reduce the hegemonic perception for positioning women's identities, especially by integration into the workforce.

4 CONCLUSION

The integrating of gender and religion in the case of Indonesia migrant workers nowadays is a part of women's social transformation and is proof of positive Muslim activities abroad. Javanese Muslim women who work in Hong Kong are still practicing their religious beliefs by conducting activities for strengthening their identity. Thus, the identity of Indonesians as one of largest communities in the world is also sustained through the female migrant workers in Hong Kong. They are not just hindered by gendered ideas of patriarchal systems which twisted their flexibility as Indonesian women but become embedded in a strong and vibrant community of Muslim representation abroad by being Indonesian migrant workers in Hong Kong.

REFERENCES

Allmark, P., & Wahyudi, I. 2019. Travel, sexuality, and female Indonesian domestic migrant workers in Hong Kong. *Continuum, 33*(5), 630–642.

Anggaunitakiranantika, A. 2016. Awakening Through Career Woman: Social Capital for Javanese Migrant Worker in Southeast Asia. In *ASEAN/Asian Academic Society International Conference Proceeding Series*.

Anggaunitakiranantika, Hamidi, M. 2020. Emotional Entanglement and Community Empowerment of Transnational Migrants' Families: A Cross-Sectional Study in Malaysia and Indonesia. *Glob Soc Welf* 7 (4), 395–404. https://doi.org/10.1007/s40609-020-00191-3

Appadurai, A. 2006. *Fear of Small Numbers*. London: Duke University Press.

Bertrand, J. 2004. *Nationalism and Ethnic Conflict in Indonesia*. UK: Cambridge University Press.

Constable, N., 2010. Telling tales of migrant workers in Hong Kong: transformations of faith, life scripts, and activism. *The Asia Pacific Journal of Anthropology*, 11(3–4), pp. 311–329.

Ford, M and Parker, L. 2008. *Women and Work in Indonesia*. London: Routledge.

Hamidi, M. 2017. *Indonesian Women and Their Unique Transnational Migration Experiences in Malaysia. Home and Away*. Kuala Lumpur: University of Malaya Press.

Ionesco, D, and Christine A. 2009. Introduction, *Gender and Labor Migration in Asia*, Switzerland: International Organization for Migration.

Kiranantika, A., 2020. Arising in Migration: Forming a Power through Connectivity for Javanese Women. *KnE Social Sciences*, pp. 312–327. DOI: https://doi.org/10.18502/kss.v4i10.7419

Koning, J & Nordic Institute of Asian Studies. 2000. *Women and households in Indonesia: cultural notions and social practices*. Surrey: Curzon Richmond

Mahler, S.J. and Pessar, P.R., 2001. Gendered geographies of power: Analyzing gender across transnational spaces. *Identities*, 7:4, 441-459, DOI: 10.1080/1070289X.2001.9962675

Miles, M.B and M. B. Huberman. 1994. *Qualitative Data Analysis: An Expanded Sourcebook (2nd Edition)*. London: Sage Publication.

Neuman, Lawrence W. 2000. S*ocial Research Methods: Qualitative and Quantitative Approaches, A Pearson Education Company*. United States of America.

Rapoport, R., Bailyn, L., Fletcher, J.K and Pruitt, B.H. 2002. *Beyond Work-Family Balance: Advancing Gender Equity and Workplace Performance*. San Francisco: Jossey Bass.

Ridgeway, C. L., and Shelley J. C. 2004. Unpacking the Gender System: A Theoretical Perspective on Gender Beliefs and Social Relations. *Gender & Society*, vol. 18, no. 4, Aug. 2004, pp. 510–531, doi:10.1177/0891243204265269.

Runnals, D. R. 2002. Gender Concept in Female Identity Development. *Women in Indonesian Society: Access, Empowerment and Opportunity*. Yogyakarta: Sunan Kalijaga Press.2002

Sheller, M. 2011. *Mobility*. Sociopedia.isa. USA. Doi: 10.1177/205684601163

Skeggs, B. 2004. *Class, Self, Culture*. London: Routledge.

Turner, V. 1974. *The Ritual Process*. London: Pelican Books.

Williams, C. P., & Widodo, A. 2009. Circulation, Encounters and Transformation: Indonesian Female Migrants. *Asian and Pacific Migration Journal*, 18(1), 123–142. https://doi.org/10.1177/011719680901800106

Community Empowerment through Research, Innovation and Open Access – Sayono et al (Eds)
© 2021 Copyright the Author(s), ISBN 978-1-032-03819-3

Innovation in *batik tulis* with ICT technology for sustainability design

F. Abdullah
Universitas Pendidikan Indonesia, Bandung, Indonesia

B.T. Wardoyo*
Universitas Trisakti, Jakarta, Indonesia

A.M. Adnan
University Technology of Mara, Selangor, Malaysia

ABSTRACT: The development of ICT technology (Information and Communication Technology) is currently very rapid, creating tools to support human life. Information and communication technology are devices consisting of hardware, software, processes, and systems used to help the communication process between humans be successful. The problem in creating *batik tulis* clothing is that there have not been many collaborations with ICT technology so far. Both ICT technology and the creation of *batik tulis* are still running, respectively. On the other hand, today's human needs for clothing are not limited to body coverings and fashion but also develop as a means of security, communication, and data storage. The method used in this paper is qualitative-experimental using the design creation stage. The finding of this creation is that human needs for batik are also increasing; it is necessary to create a creation that combines ICT technology and *batik tulis* to answer human needs in the era of 4.0. This creation can be a solution to human needs for security, communication, and data storage that is integrated into a *batik tulis* cloth.

Keywords: *batik tulis*, ICT, sustainability design

1 INTRODUCTION

The industrial revolution 4.0 is part of globalization. The era of Industrial Revolution 4.0 has fundamentally caused many changes way of thinking of humanity, including the way of life and the way of relating one human being to another. Era 4.0 has disrupted all human activities in various aspects of life, not only in technology but also in the economic, social, and political fields (Prasetyo & Trisyanti 2018). Globalization then entered a new stage through the paradigm of The Fourth Industrial Revolution, which states that the world has entered four phases of the revolution, namely: (1) Industrial Revolution 1.0 through the invention of the steam engine in the 18th century; (2) Industrial Revolution 2.0 in the form of electricity usage in the 19th century; (3) Industrial Revolution 3.0 in the form of computerization in the 1970s, and; (4) Industrial Revolution 4.0 through artificial intelligence engineering and IoT (Internet of Things) as the backbone of human-machine progress and connectivity (Schwab 2017).

The rapid progress of ICT (Information and Communication Technology) in the Industrial Revolution 4.0 era has changed the face of the world with the consequences that followed. The era of the Industrial Revolution 4.0 in the form of sophisticated information and communication technology has encouraged the exchange of information and communication very quickly without the constraints of time and space boundaries (Dryden & Voss 1999). The Agency for Science and

*Corresponding author: bambangtri@trisakti.ac.id

DOI 10.1201/9781003189206-16

Culture of the Nations (UNESCO) in 2002 stated that the integration of ICT into the world of education, especially into the human learning process, has three main objectives, namely: (1) to build knowledge-based community behavior (knowledge-based society habits); (2) to develop ICT literacy skills; (3) to increase the effectiveness and efficiency of the learning process (Mawardi 2012). UNESCO's support explains the stronger push for ICT integration into other cultural artifacts, including *batik*.

The use of ICT for collaboration in conventional batik products is currently still minimal. For this reason, human creativity is needed to collaborate with technology and *batik* fashion traditions. Creativity can be defined as an individual's capacity to develop ideas based on divergent ways of thinking, which is higher than convergent ways of thinking. Creativity can also be interpreted as how an individual is more concerned with solving problems in the real-world context (Basadur & Gelade 2014). The creative process in terms of answering a problem (problem solver) and this utilization process produces prototypes that can be worn by a broad audience.

Batik artifacts are often positioned as traditional products produced by local craftsmen but become the cultural image of a nation. *Batik tulis* also faces the threat of mass production from printed (filter print) fabrics claiming to be batik. The existence of *batik tulis* in aesthetics and commercial art is part of the demand and supply conditions in the market. Various cultural protections, in this case, batik motifs, have been carried out, including learning local content in schools even though the implementation of batik as cultural learning is carried out unevenly in various regions in Indonesia (Farid 2012). This makes Batik protection very critical from an industrial perspective. Poon's research results found that mass-produced batik is required to be adaptive to current clothing needs (Poon 2017).

Batik tulis clothing is closely related to the characteristics of the user. In general, the users of batik tulis are a certain level of society (generally the upper economy), have excellent economic capacity, and need products that elevate their personality. According to Gaspersz's (2002) research, several aspects need to be considered to achieve product quality that satisfies users. These aspects include: (1) Performance is a function or use; (2) Additional characteristics or features, namely secondary or complementary characteristics; (3) reliability (reliability); (4) Compliance with specifications; (5) Durability; (6) Convenience (serviceability); and (7) Aesthetics (Gaspersz 2002, in Mandegani et al. 2018). Then a piece of batik is also required to be adaptive to user satisfaction, especially *batik tulis* users and the phenomenon of the Industrial Revolution 4.0. This paper seeks to investigate to what extent is the collaborative adaptive ability between the batik tradition and the ICT technology.

2 METHODS

The creation process to utilize ICT in creative batik in this era of 4.0 used a descriptive - experimental approach. Experimental creation research provides an opportunity for designers not to be confined to a social research framework because knowledge in academia is the knowledge that comes from practical experiences carried out by the creator (Murwanti 2016). The initial stage in utilizing ICT is identifying needs, then the definition of needs is carried out, followed by ideas that can be poured out, then making a prototype (Figure 1).

In the process (Figure 1), there are several parts to definite identification, namely what aspects of the research object's problem. At the definition stage, which is the stage of narrowing down the goals of creating ICT use. The ideas stage is when the design process finds new ideas. Likewise, at

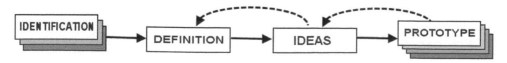

Figure 1. Stages of the ICT utilization process.

the prototype stage, namely making modeling, final predesign. Henceforth, Figureure 1 is presented in the discussion.

3 RESULTS AND DISCUSSION

Batik was chosen in the creation of prototypes because it is inherent in Indonesian culture. Batik has been designated as an intangible cultural heritage from Indonesia by UNESCO since October 2, 2009 (Iedarwati 2017; Kharismawati 2017). The variants of batik motifs in objects as inventions are very diverse, including gifts or souvenirs, even though they are not yet the leading choice for Japanese people to make batik as souvenirs (Kharismawati 2017). This is a challenge for batik to provide added value in its use. In this prototype, a batik motif inherent as a characteristic of Indonesian regions will be selected, namely, the upper clothing, which is attached to its function, namely as body armor, but this top clothing is often used as a gift souvenir.

In the prototype selection, the Mega Mendung motif was chosen, which is the hallmark of *Batik* Cirebon, West Java. This motif is a cloud shape. This motif is very familiar as a characteristic of the West Java motif, which symbolizes the seven soil layers. This batik is batik in the coastal areas of Indonesia so that the cloud line motifs are more masculine and suitable for use in men's batik, although in the future, this prototype can be widely used in women's clothing.

The use of technology, in this case, is inspired by a series of current works that cannot be separated in various technologies ranging from smartphones, power banks, data storage (data storage) so that these three objects have become objects that cannot be separated or attached to the context of everyday life. Various professions, especially men, will not be separated in these three objects and even work with these objects. It is not surprising that the creation of batik is attached to the various technologies that humans use in their work today. Of course, the character of the technology is simple and easy to put back on so that technology devices can be removed and put on different clothes like the clothes we know today. The design stages are as follows:

The initial stage in the creation process is identification (Figure 1). At this stage, identification of the problems that exist in the creation of batik so far is carried out. Obtained identification of *batik tulis* products so far is still a lack of collaboration in utilizing the ICT in *batik tulis* clothing products. In addition, departing from the user aspect, especially batik with high economic value, is still often troublesome by various technologies that must be handheld users in every activity such as smartphones, power storage, and storage data. Identification also shows that users also need a sense of security when doing activities, so technology such as a location scanner (Global Positioning System) can be applied to create batik.

The next stage is the definition, namely the determination of the purpose of creation. The definition at this stage is to create the use of ICT that can support users in the documentation, data storage, security, and communication. The definition of giving added value is the combination of ICT with *batik tulis* clothing. The definition includes the possibility of creating a *batik tulis* product that can create novelty for local artisans. It is hoped that this novelty will be able to revive and increase local *batik*raftsmen's productivity. The third stage is an idea or idea, which generates ideas that can be done by utilizing current ICT technology. The idea offered is the use of combining ICT in batik. This stage of the idea also adapts to the ability of batik to store ICT devices.

In the design sketch (Figure 2) above, there is an idea that creates the prototype. As with *batik tulis* in general, the *batik* clothes above are designed for specific uses, not everyday use. It is also necessary to consider the ease of removing ICT devices when washing or dry cleaning is done, so Velcro adhesive rubber can be used to attach ICT technology to the batik clothing. The treatment of *batik tulis* itself requires special treatment that distinguishes it from ordinary clothing made from printing.

The final stage of this creation process is a prototype. The prototype or archetype is the first form and can also be called the initial form. In design, prototypes are made before being developed into a final product or as part of development before being mass-produced. The prototype stage is an essential activity in the process of creating new products. The purpose of the prototype is

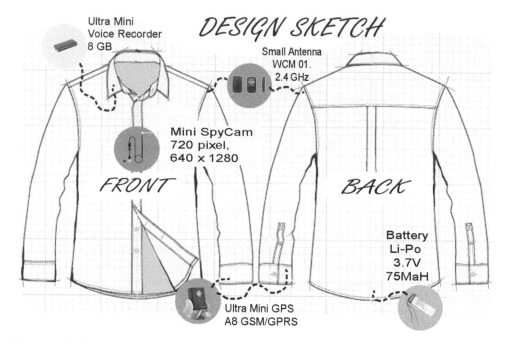

Figure 2.   Design sketch as an initial design idea.

Figure 3.   The prototype of using ICT in *batik tulis*.

to explore new possibilities or answer pre-existing problems (Elverum et al. 2016). The prototype stage is an essential part of the process of utilizing ICT in *batik tulis* clothing.

In this utilization process, it is also very possible for sustainable design to occur, from prototypes to ideas, likewise, from the idea stage back to redefinition. This reprocessing process aims to improve the use of ICT in written batik as part of product improvement.

Figure 3 portrays a prototype using ICT in *batik tulis* clothing. As with other prototypes, there are other alternatives, especially in batik motifs and embedded ICT technology. The incorporation of ICT technology in *batik tulis* clothing can adapt to the user's different and situational needs.

4 CONCLUSION

The progress of the times and technology at this time is speedy and cannot be avoided. The options available in responding to the progress of the times and technology are adaptively following it. This adaptive ability can save batik artifacts as part of cultural products. This adaptive ability is essential because the batik user community's demands are also growing, not just covering the body and fashion. The design of ICT use in the *batik tulis* clothing prototype seeks to highlight additional features, performance, and aesthetics.

Utilization of ICT is beneficial for batik's life force. Support for the advancement of ICT in the Industrial Revolution 4.0, which is increasingly progressing rapidly and sophisticated, should be utilized by batik actors, designers, craftsmen, marketers, and all stakeholders who benefit from *batik* artifacts as cultural products. The use of this ICT can make batik more and more carried away with the development of the times' spirit (zeitgeist). Thus, in the future era of the Industrial Revolution 5.0, several discoveries such as the ability to change the color of *batik* clothes automatically, the ability to self-clean on batik (which is being/has been done by the Center for *Batik*rafts, Yogyakarta), to the ability of automatic batik to be bacteria and viruses, of course very welcomed by the world community.

The creation of batik in this article, which is sustainable design and supports security in the form of being connected by GPS (Global Positioning System), storing data, carrying out/record communications, is still a prototype. The downside of creating this creation is that it must be removed when it is about to be cleaned in the laundry. The technology attached to the prototype on the *batik tulis* clothing is still not ready for the wet washing treatment. However, collaborative researchers believe that this creation's idea can be realized in a short time and can be used by people around the world.

REFERENCES

Basadur, M. Gelade, G. 2014. Creative Problem-Solving Process Styles, Cognitive Work Demands, and Organizational Adaptability. *Journal of Applied Behavioral Science*, 50(1), 80–115.

Dryden, G., Voss, J. 1999. *The Learning Revolution : To Change the Way the World Learn*, The Learning web, Torrence, US. http:www.thelearningweb.net.

Farid, Muhammad Nur. 2012. Peranan Muatan Lokal Materi *Batik tulis* Lasem Sebagai Bentuk Peles-tarian Budaya Lokal. *Komunitas*. Vol 4.1.2012. 90–121.

Gaspersz, V. 2002. *Pedoman Implementasi Program Six Sigma*. Jakarta: Gramedia Pustaka Utama.

https://ich.unesco.org/en/RL/indonesian-batik-00170, downloaded on 9th September 2020

Iedarwati, Pradewi. 2017. Entrepreneurial Characteristics of Betawi Batik women Activists. *Journal of Entrepreneur and Entrepreneurship*, Vol. 6, No. 1, March 2017, 17–24. ISSN 2302-1802 print / ISSN 2580-9393 online.

Kharismawati, Mery. 2017. Baju Batik sebagai Omiyage: Studi Kasus Pada Mahasiswa Jepang Yang Pernah Belajar di UGM Tahun 20019-2017. *Izumi*, Volume 6, No 2, 2017e-ISSN: 502-3535, p-ISSN: 2338-249X

Mandegani, G.B., Setiawan, J., Haerudin, A., Atika, V. 2018. Persepsi Kualitas *Batik tulis*. Dinamika Kerajinan dan Batik. *Majalah Ilmiah*, 35(2), 75–84.

Mawardi, I. 2012. ICT (Information and Communication Technology) sebagai Wahana Transformasi Pendidikan. *Tarbiyatuna; Jurnal Penelitian dan Pendidikan Islam*, 3(1), 99–114.

Murwanti, A. 2016. Penciptaan Desain Berbasis Praktik Eksperimental Sebagai Penelitian Ilmiah. *UL-TIMART Jurnal Komunikasi Visual*, 6(2), 7–18.

Poon, S.T.F. 2017. The Journey to Revival: Thriving Revolutionary Batik Design and Its Potential in Contemporary Lifestyle and Fashion. *International Journal of History and Cultural Studies* (IJHCS), 3(1), 48–59.

Prasetyaningtyas. 2011. *Perkembangan Motif dan Warna Batik Mega Mendung di Kawasan Sentra Batik Trusmi, Cirebon, Jawa Barat*. Skripsi, Program Studi Pendidikan Seni Kerajinan, Fakultas Bahasa dan Seni, Universitas Negeri Yogyakarta.

Prasetyo, B., Trisyanti, U. 2018. Revolusi Industri 4.0 dan Tantangan Perubahan Sosial. *Prosiding SE-MATEKSOS 3*, Strategi Pembangunan Nasional Menghadapi Revolusi Industri 4.0.

Rosyadi, S. 2018. *Revolusi Industri 4.0: Peluang dan Tantangan Bagi Alumni Universitas Terbuka*. https://www.researchgate.net/publication/324220813_REVOLUSI_ INDUSTRI _ 40 # fullTextFileContent

Schwab, K. 2017. *The Fourth Industrial Revolution*. New York: Crown Business.

Community Empowerment through Research, Innovation and Open Access – Sayono et al (Eds)
© 2021 Copyright the Author(s), ISBN 978-1-032-03819-3

The influence of administrative literacy on employee's performance on the perspective of gender among local government administration staff

A. Winarno* & Zulaikah
Universitas Negeri Malang, Malang, Indonesia

ABSTRACT: This research aims to examine and describe the degree of administrative literacy of office staff, which includes their knowledge, skills, and abilities, as well as the influence of their literacy on their performance. Additionally, this paper examines the perspective of gender as a contributing factor to performance determination. This research employed a quantitative research approach by using expose facto causality. The data was collected through questionnaire distribution, surveys, and interviews, while the data analysis was performed by using regression analysis and chi-square analysis. The research results indicate that administration knowledge carries no effect on employee performance. In contrast, their administration skills and abilities influence the performance of employees.

Keywords: gender, administrative literacy, employee's performance

1 INTRODUCTION

In these current years, the administrative service has also experienced a transformation. It starts from the role and responsibilities of the secretary or administrator, which are commonly known as administrator functions and tasks. This function has changed into an instrument or media used during the administration process to support office administration activity. As supported by an argument of Grönlund et al. (2007), the needs of skills and abilities and competence in office administration are now related to the electronic administration service. A secretary or administrator is expected to be able to provide a service based on the development of current technology to perform the task and function better. A secretary's primary role is not only limited to typing, writing, and preparing documents, but the secretary also needs to be able to perform any other assignments and functions related to administration (Laswitarni 2017).

The office administrator or secretary plays an essential role in almost every office business and matter. In addition, the board of directors of an office highly rely on the presence and function of the administrator or secretary. Almost 80 percent of the board of directors' assignments and tasks are handled by the administrator or secretary. Siregar (2018) states that the flow and circulation of the board of directors' duties highly depend on the administrator's performance. Thus, the office administrator serves as administrative support of the board of directors in the office.

Additionally, the transformation of a secretary career from a professional into an administrative professional, as suggested by IAAP, aims to better recognize the secretary's profession. Consequently, this change demands an individual meet certain criteria. One of these criteria is administrative literacy. Administrative literacy is vital to supporting professional administrative works these days, especially in organizing a company. Winarno, et al. (2019) argue that administrative literacy is an essential ability of the administrative professional. Administrative literacy includes the ability to manage assignments and tasks, the understanding of procedure and organization policy, the ability to communicate and the ability to cooperate.

By developing administrative literacy, an administrative professional is expected to perform the assignments and contribute to the company better. Besides, an administrative professional is

*Corresponding author: agung.winarno.fe@um.ac.id

DOI 10.1201/9781003189206-17

demanded to be able to understand the needs of the client to provide proper service, think critically, and perform in accordance with the applied ethics code (Salam 2015).

The 21st Century is commonly known as the era of knowledge. Almost all aspects of the 21st century are closely related to knowledge. Moreover, P21 or "Partnership for 21st Century Learning" has developed a 21st century learning framework that demands learners acquire abilities, knowledge, and understanding about technology, media, and information. Also, the learners are expected to acquire life skills to survive in current society (Wijaya et al. 2016).

With such a variety of duties and responsibilities that must be carried out, human resources should also have high competencies for maximum performance. It is in line with the opinion of Salam (2015) that an organization requires reliable and competent human resources in their fields to achieve the goals or objectives. When an organization employs an individual in accordance with their fields and has competence in that field, the organization gains optimal performance and vice versa (Juned et al. 2016). Meanwhile, Winarno (2016) revealed that education and skills are considered capable of improving employees' quality or performance.

The quality of performance is highly affected by the characteristics of each employee. One of these characteristics is gender. This characteristic is able to influence the way a person works. The current secretary or administrator profession has a balanced composition between men and women with no different treatment. The division of a position is not carried out by gender but is based on performance. In line with the opinion of Jamilah, et al. (2007), gender is presumed to be one of the factors that influence performance and changes in task complexity and compliance with ethics. It is inversely proportional to the opinion of Sabrina, et al. (2016), who states that jobs in the administration sector are more likely to be dominated by women because women are perceived as more capable or more competent. In other words, women's knowledge of administration is assumed to be broader and more skilled in terms of services, even though research results suggest that gender has no connection with employee performance (Ramadhani & Adhariani, 2015). In fact, research finds that literacy has an effect on performance (Winarno & Wijijayanti, 2018). Likewise, identical findings form Anggraeni (2011) confirmed that knowledge, skills, and attitudes are determinants of performance. Thus, if employees have good administrative literacy, the resulting performance will also be optimal to achieving organizational goals. Based on the description above, the researcher tries to determine how administrative literacy influences employee performance in regional government offices from a gender perspective.

2 METHODS

This study employed a descriptive quantitative explanatory research approach with a survey method, while the data collection was carried out using questionnaires and interviews. The independent variable included knowledge, skills, and abilities, while the dependent variable was employee performance. The knowledge variable was viewed from four aspects, namely a specific understanding of the scope of work, the way the work was carried out, the suitability of the knowledge possessed with the field of work and understanding in facing job challenges. The skills variables included the ability to plan, accuracy in work, the ability to work in groups and creativity. The ability variables were intellectual abilities, physical abilities, employee performance variables seen from the aspects of effectiveness, efficiency, quality, accuracy, productivity, safety and security.

The population in this study were all employees at the Regional Secretariat of Blitar City. Samples were taken with a saturated sampling technique, resulting in 71 people. The data analysis used multiple linear regression in determining the relationship between the effects of variables. Meanwhile, the chi-square difference test ($\lambda 2$) was used to test whether there were differences in knowledge, skills, abilities, and performance between men and women.

3 RESULTS AND DISCUSSION

The research results on the level of knowledge show that the knowledge of employees in the administration sector obtained a 3.92 score. This is considered good within the category with an interval scale in the range of 3.41 to 4.20. The educational background (formal and non-formal) of

Table 1. The results of multiple regression analysis.

Coefficients[a]

Model	Unstandardized Coefficients		Standardized Coefficients	t	Sig.
	B	Std. Error	Beta		
1 (Constant)	18.956	5.668		3.344	.001
Knowledge	.054	.187	.031	.287	.775
Skills	.528	.156	.381	3.376	.001
Abilities	.407	.186	.241	2.186	.032

Table 2. The comparison of performance between male and female administration professionals.

Gender * Performance Cross tabulation

Gender		Performance			Total
		Moderate	Good	Very Good	
Male	Count	6	28	1	35
	% within Performance	60.0%	49.1%	25.0%	49.3%
Female	Count	4	29	3	36
	% within Performance	40.0%	50.9%	75.0%	50.7%
Total	Count	10	57	4	71
	% within Performance	100.0%	100.0%	100.0%	100.0%

Table 3. The results of chi-square analysis on the variable of performance.

Chi-Square Tests

	Value	Df	Asymp. Sig. (2-sided)
Pearson Chi-Square	1.404[a]	2	.496
Likelihood Ratio	1.453	2	.484
Linear-by-Linear Association	1.219	1	.270
N of Valid Cases	71		

a. 3 cells (50.0%) have an expected count of less than 5. The minimum expected count is 1.97.

employees shows the highest number. It describes their educational background suitability with their field of work. In contrast, the lowest score is observed in understanding the concept of equipment oriented procedures (sending and receiving e-mails, voice mails, using Local Area Networks, online databases, and understand the use of multimedia). Employee skills in the administrative field obtained a score of 4.07, which is in the good category. The average answer of the respondents with the highest score is the skills in communicating with others and preference to work in teams, while the lowest score is related to the willingness to work beyond the target. The ability of employees in the administrative sector obtains a score of 3.90, which is classified in the good category, the highest score is on the item of accuracy, thoroughness, and critical thinking, while the lowest score is in the type of work that is related to the need for accuracy with numbers. The performance of employees in the administration sector attains a score of 3.90 which was in the good category, while the highest score is observed in the aspect of the ability to use office equipment and supplies while the lowest score is in the aspect of laziness to work with target pressure (Table 1).

Based on the results of the data in Table 1, the variables that have an effect and are significant are the Skills (X2) and Abilities (X3) variables, while the Knowledge variable (X1) has no significant impact. The analysis results on the variable of performance in administration professional are presented in the following Table 2.

From Table 3, it is found that the significance level of performance in the administration sector is 0.496. This value is greater than the significance level $(\alpha)0.496 > 0.05$. Thus, Ho is accepted, indicating no difference in the administrative sector's performance level between male employees and female employees at the Regional Secretariat of Blitar City.

In addition, knowledge in the field of administration also carries no effect on employee performance. This can be due to the lack of knowledge of employees in carrying out equipment-oriented procedures. Meanwhile, this knowledge is vital for administrative professionals since it includes knowledge of procedures for sending and receiving e-mail, voice mails, using Local Area Networks, online databases and procedures regarding the usage of multimedia. The average working period of employees is 11-15 years, where at the beginning of work, employees are not used to using technology-based equipment, which is now increasingly sophisticated. However, on average, employees have an educational background that is following their current work field. Thus, they only need to improve their knowledge of currently developing technologies to widen their knowledge, with no broad administrative insight, their performance declines, and vice versa.

Based on open questionnaires that have been distributed by researchers, it shows that the employee's performance increases if supported by existing facilities and infrastructure. Complete and useful facilities and infrastructure improve the employees' excitement to carry out the given tasks and responsibilities. However, some of the existing facilities and infrastructure seem incomplete, and while some should be replaced, often no action has been taken. Consequently, employees feel less comfortable in carrying out their duties; for instance, when the weather is hot, and AC cannot be used. Thus, it is necessary to modernize existing facilities and infrastructure to make the employees work comfortably. Furthermore, later, it will improve organizational performance. This is in line with the statement Hartono (2014) that office facilities and infrastructure are intended to support the smooth work process and facilitate internal and external communication to fulfill employee welfare. Finally, it will improve organizational performance.

These study results are in accordance with research conducted by (Ardiana & Brahmayanti, 2010), which stated that knowledge does not show a significant effect on employee performance. A similar study conducted by Laoh (2016) also revealed that knowledge management has no significant impact on employee performance.

Based on the research results, administrative skills have a positive and significant effect on employee performance. In other words, employees with high administrative skills result in high performance and vice versa. Based on the research results, employees' skills are in a good category, especially the skills in communicating with others. With these skills, employees will easily interact with others. Good interaction will lead to good cooperation. When employees are required to work with the team, the employees will be able to do it well thus, the work will be easier to complete, and the results are also maximized. This means that when employee skills improve, they will be able to improve their performance.

The results of this study are in accordance with several previous studies, such as research conducted by Fadhil (2016), which showed that skills have a positive and significant effect on performance. Research conducted by Makawi et al (2015) stated that skills have the most dominant and significant influence on employee performance compared to other variables. Similar research conducted by Kartika & Sugiarto (2016) showed that competency consists of skills indicators where the results positively and significantly affect employee performance. Research carried out by Ataunur & Ariyanto (2015) also stated that the skill variable has a more dominant and significant influence on performance than other variables. Another study shows that skills have a reasonably dominant or essential influence on performance. Thus, skills definitely lead an organization to achieve its goals. Therefore, it is necessary to increase employees' skills. Employees with excellent skills become more proficient in carrying out work, specifically in the administrative field.

This research result confirms that the ability in the field of administration has a positive and significant effect on employee performance. Employees with high administrative abilities bring outstanding performance and vice versa. Capacity in the field of administration represents the abilities of an employee in carrying out his/her duties and responsibilities, especially in the administrative field. The research findings show that employee ability in the administrative field is in the good category where the average employee can carry out administrative work. If a problem occurs, they are able to solve it well. Administrative activities related to numbers and calculations can make employees become careful, thorough, and able to think critically; thus, the resulting performance is also more optimal. The more frequently this ability is used, good abilities increase, along with good performance and vice versa. Therefore, it is important for organizations to improve the ability of their employees, especially related to the ability to count.

These study results are in accordance with several previous studies, such as research conducted by Suhartini (2015), which showed that employee ability positively and significantly affects their performance. Research conducted by Anggraeni (2011) mentioned that there is a positive and significant influence on employee performance. Similar research conducted by Sriwidodo & Haryanto (2010) showed that the ability in the competency variable has a positive and significant effect on employee performance. Additionally, Jaya (2012) revealed a positive and significant influence on employee performance and ability. This study also shows that ability has a positive and significant effect on performance.

The research result shows that the level of knowledge, skills, abilities, and performance in the administration sector has no significant difference between male and female employees. This means that men and women have the same level of knowledge, skills, abilities, and performance in the administrative field. Men have higher knowledge, while women have much higher in skills or abilities.

The performance between men and women is also balanced due to the knowledge, skills, and abilities of the Blitar City Regional Secretariat employees are in the excellent category. Therefore, the gender difference has no effect on employee performance. The difference in employee knowledge is not seen from gender but from their educational background or their working experience. However, educational background is not a standalone factor that leads to the increasing of knowledge. Likewise, employee skills and abilities differ not because of gender but based on the training or technical guidance that employees have taken. The frequency of training improves their skill. This is in accordance with the opinion of Jaya (2012) that skills are behaviors related to tasks that can be mastered through learning and can be improved through training and assistance from others.

4 CONCLUSION

The results showed that local government employees' literacy with an average working period of over 11 years indicates a good level, as reflected in their level of knowledge, skills, and abilities in the field of administration, including their performance. Administrative knowledge does not have a significant effect on employee performance, whereas administrative skills and abilities have a significant effect on employee performance. The gender aspect is not a factor that affects differences in knowledge, skills, abilities and performance.

REFERENCES

Anggraeni, N. 2011. Pengaruh Kemampuan dan Motivasi Terhadap Kinerja Pegawai pada Sekolah Tinggi Seni Indonesia (STSI) Bandung Oleh: Nenny Anggraeni. *Jurnal Penelitian Pendidikan* Vol. 12.

Ardiana, I.D.K.R.I.A. Brahmayanti, S. 2010. Kompetensi SDM UKM dan Pengaruhnya Terhadap Kinerja UKM di Surabaya. *Jurnal Manajemen Dan Wirausaha.* https://doi.org/10.9744/jmk.12.1.pp.42–55.

Ataunur, I., & Ariyanto, E. 2015. *Pengaruh Kompetensi dan Pelatihan terhadap Kinerja Karyawan PT. Adaro Energy Tbk.* Telaah Bisnis.

Fadhil, M. 2016. Pengaruh Kompetensi Sumber Daya Manusia terhadap Kinerja Pegawai pada Balai Latihan Kerja Industri Makassar. *Perspektif: Jurnal Pengembangan Sumber Daya Insani,* 1(01), 70–81. https://doi.org/https://doi.org/10.26618/perspektif.v1i1.155

Grönlund, Å., Hatakka, M., & Ask, A. 2007. Inclusion in the e-service society - Investigating administrative literacy requirements for using e-services. In Lecture Notes in Computer Science (including subseries

Lecture Notes in Artificial Intelligence and Lecture Notes in Bioinformatics). https://doi.org/10.1007/978-3-540-74444-3_19

Hartono, D. 2014. Pengaruh Sarana Prasarana dan Lingkungan Kerja Terhadap Kinerja Pegawai Dinas Pendidikan Kota Banjarbaru. *Journal Kindai,* 10(02), 142–155.

Jamilah, S., Fanani, Z., & Chandr, G. 2007. Pengaruh Gender, Tekanan Ketaatan, dan Kompleksitas Tugas terhadap Audit Judgment. *Simposium Nasional Akuntansi* 10.

Jaya, I. 2012. Pengaruh Kemampuan Dan Motivasi Kerja Terhadap Kinerja Pegawai Dinas Pendidikan Kabupaten Tanjung Jabung Barat. *Jurnal Penelitian Universitas Jambi.*

Juned, A., Jonathan, L. R., & Lau, E. A. 2016. Pengaruh Disiplin, Kompetensi dan Kepemimpinan Terhadap Kinerja Pegawai Dinas Tenaga Kerja Kota Samarinda. *Ekonomia.*

Kartika, L. N., & Sugiarto, A. 2016. Pengaruh Tingkat Kompetensi Terhadap Kinerja Pegawai Administrasi Perkantoran. *Jurnal Ekonomi Dan Bisnis.* https://doi.org/10.24914/jeb.v17i1.240

Laoh, C. F. 2016. Pengaruh Manajemen Pengetahuan, Keterampilan dan Sikap Kerja Terhadap Kinerja Pegawai. *Jurnal Berkala Ilmiah Efisiensi,* 16(04).

Laswitarni, N. K. 2017. Kiat Pengembangan Sekretaris Profesional. In Forum Manajemen STIMI Handayani Denpasar (pp. 9–14).

Makawi, U., Normajatun, & Haliq, A. 2015. *Analisis Pengaruh Kompetensi Terhadap Kinerja Pegawai Dinas Perindustrian Dan Perdagangan Kota Banjarmasin.* Al-Ulum Ilmu Sosial Dan Humaniora.

Ramadhani, Z. I., & Adhariani, D. 2015. Pengaruh Keberagaman Gender Terhadap Kinerja Keuangan Perusahaan dan Efisiensi Investasi. *Simposium Nasional Akuntansi* XVIII.

Sabrina, T., Ratnawati, R., &Setyowati, E. 2016. Pengaruh Peran Gender, Masculine Dan Feminine Gender Role Stress Pada Tenaga Administrasi Universitas Brawijaya. *Indonesian Journal of Women's Studies,* 4(01). Retrieved from http://ijws.ub.ac.id/index.php/ijws/article/view/111

Salam, R. 2015. Penerapan Fungsi Administrasi Perkantoran Modern berbasis Daya Saing Organisasi dalam menyongsong MEA 2015. In SEMINAR NASIONAL "Revolusi Mental dan Kemandirian Bangsa Melalui Pendidikan Ilmu-Ilmu Sosial dalam Menghadapi MEA 2015" Himpunan Sarjana Pendidikan Ilmu-Ilmu Sosial Indonesia.

Siregar, Y. B. 2018. The Changing Roles Of Administrative Professionals In The Office Of The Future. *Jurnal Administrasi Dan Kesekretarisan,* 3(01), 29–39. Retrieved from http://jurnal.stiks-tarakanita.ac.id/index.php/JAK/article /view /136

Sriwidodo, U., & Haryanto, A. B. 2010. Pengaruh Kompetensi, Motivasi, Komunikasi Dan Kesejahteraan Terhadap Kinerja Pegawai Dinas Pendidikan. In Manajemen Sumberdaya Manusia.

Suhartini, Y. 2015. Pengaruh Pengetahuan, Keterampilan dan Kemampuan Karyawan terhadap Kinerja Karyawan (Studi pada IndustriKerajinanKulit di Manding, Bantul, Yogyakarta). *Akmenika: Jurnal Akuntansi Dan Manajemen,* 12(02). Retrieved from http://ojs.upy.ac.id/ojs/index.php/akm/article/vi ew/123

Wijaya, E. Y., Sudjimat, D. A., & Nyoto, A. 2016. *Transformasi Pendidikan Abad 21 Sebagai Tuntutan Pengembangan Sumber Daya Manusia di Era Global.* Prosiding Seminar Nasional Pendidikan Matematika.

Winarno, A. 2016. Entrepreneurship Education in Vocational Schools: Characteristics of Teachers, Schools and Risk Implementation of the Curriculum 2013 in Indonesia. *Journal of Education and Practice.*

Winarno, A., Novitasari, E. & Firdaus, R. 2019. Hubungan Administrative Literacy, Kompetensi dan Masa Kerja terhadap Kinerja Pegawai Pemerintah Daerah. In E. Pratikto, Heri, Hurriyati, Ratih, Suhartanto (Ed.), Pendidikan, Bisnis, dan ManajemenMenyongsong Era Society 5.0 (pp. 20–31). Malang: Bagaskara Media.

Winarno, A., & Wijijayanti, T. 2018. Does entrepreneurial literacy correlate to the small-medium enterprises performance in Batu East Java? *Academy of Entrepreneurship Journal.*

Community Empowerment through Research, Innovation and Open Access – Sayono et al (Eds)
© 2021 Copyright the Author(s), ISBN 978-1-032-03819-3

Impact of ecological landscape changes toward community life in Southern Malang during the 19th–20th Century

L. Ayundasari*, J. Sayono & S.D. Utari
Universitas Negeri Malang, Malang, Indonesia

ABSTRACT: Southeast Malang is one of the most dynamic areas in Malang in the 19th–20th century. This is due to the fundamental socio-economic changes from a subsistence society to an industrial society and into a working-class society. This research aims to explore the impact of environmental changes on the patterns of community life that occur there. The locations of this research are Dampit, Tirtoyudo, and Ampelgading. The research method used was historical. Data was collected through interviews, field observations, and documentation. The study results indicate the presence of social changes caused by ecological transformation as an indirect impact of dozens of private coffee plantations in this region.

Keywords: ecological landscape, environmental changes, Southern Malang

1 INTRODUCTION

Life is a dynamic process and change is a certainty, including economic field. One of the figures who studied economic change was Rostow who stated that the stages of economic development consisted of 5 phases, namely: The Traditional Society, The Preconditions for Take-off, Take-off, The Drive to Maturity, and The Age of High Mass Consumption (Rostow 1960). Economic development supported by advances in technology and investment have had a major impact on social, cultural and environmental changes such as what happened to traditional communities in remote areas in Kalimantan, Sumatra and Papua. (Petrenko et al. 2016; Hidayah et al. 2016; Pirard, et al. 2017; Imadudin 2014). A similar phenomenon also occurs in developed countries and other developing countries such as Spain, Bangladesh and Ethiopia (Watson et al. 2014; Ghosh et al. 2020; Nigussie et al. 2021). These events have a similar pattern to the development of a plantation technology-based economy in Indonesia in the 19th century.

Economic liberalism, which was marked by the freedom of foreign investment and the development of large industries, occurred in Indonesia towards the end of the 19th century – during the enactment of the liberal economic system established by the Dutch East Indies Government. At this time, liberalism was not just an invitation but took the form of a policy by giving permits to foreign private companies to invest in cash crops such as coffee, rubber, quinine, tea, sugar cane, and others (Kadir 2018). In terms of Rostow's theory of economic development, at that time, Indonesia was classified in the category of a take-off society because it had entered the industrial sector with a high growth rate, along with an established political, social and institutional framework. However, Indonesia was only a colony whose economic and political stability depended on the mother country. In addition, the Indonesian people's position was only as plantation workers or factory employees (Hudiyanto 2015).

A series of significant depression-era events, World War II, the Japanese occupation and the independence revolution, all resulted in the beginning of the collapse of Indonesia's plantation industry. This occured throughout the region, including in the former *Afdeeling* Malang in Malang's

---

*Corresponding author: lutfiah.fis@um.ac.id

DOI 10.1201/9781003189206-18

coffee plantation centers of Dampit and Ampelgading. The two districts that had been successful in coffee production for almost 50 years were slowly dying out. This is due to many factors, including the international failure of plantation products due to war, the absence of a support system for the plantation product care and processing, such as *Mantri* who usually handle and provide counseling, and the insufficient attention of the new government to the plantation industry. During the Old Order and New Order governments, this region was marginalized. The low level of welfare caused the new generation in this region to choose to work abroad. Until now, this area has been known as a migrant worker enclave. Therefore, this study aims to find out details about the effects of changes in the ecological landscape as a result of economic development on the community life pattern changes in Southeast Malang in the 19th–20th centuries.

2 METHODS

The method used in this study was a historical research method, which consists of four stages: heuristics, criticism, interpretation, and historiography (Kuntowijoyo 2005). Historical sources used in this study consisted of two types, namely primary and secondary. The primary source of this research was the archives on plantations in Ampelgading and Dampit. Meanwhile, secondary sources were obtained from reports and interviews with historical actors. Observation activities were carried out by following evidence of change in the ecological landscape during the period of large plantations, such as coffee plantations, factories and water management for coffee production. All the collected data was analyzed for its validity through internal and external criticism processes. The next stage was the interpretation, which consisted of interpreting and analyzing data using sociological and economical lenses so that the historical narrative presented in narrative form fully described the phenomenon. The final stage of this research was historiographical, consisting of the examination of of changes in the ecological landscape on the people's patterns of life in Southeast Malang in the 19th century.

3 RESULTS AND DISCUSSION

3.1 *Geographical and administrative conditions of Southeast Malang*

Southeast Malang is a term deliberately used to refer to three sub-districts, namely Dampit, Tirtoyudo, and Ampelgading. This term was developed to make it easier to refer to the region. The three districts are geographically located on the slopes of Mount Semeru, with land contours consisting of valleys, slopes and hills. There are two types of soil that dominate this area, namely latosol reddish-brown and mediteran reddish brown. The land is famous for being fertile and very suitable for use as agricultural and plantation land. This was the main motivation for the large-scale clearing of plantation land by the Dutch government in the 19th century, both by the private and Dutch colonial governments. Apart from fertile land, this region also has an abundant supply of water due to the dense forest nearby. One of the upstream rivers in this area is the Lesti River. Besides that, there are also dozens of water sources that are spread across the three sub-districts.

The perfect mix of soil types and water sources in Southeast Malang has invited many newcomers to settle in this region. However, before the big plantation era in the mid-19th century, this area did not have many inhabitants for several reasons, including the contours of the hilly land that limited access, most of which were still wild and needed further processing. In addition, there were no production plants to be used for the fulfillment of living needs. The remoteness of this area led to there being no practical modes of transportation. Apart from this, East Java's population distribution until the beginning of the 19th century was not evenly distributed and remains as dense as it is today. At that time, the population distribution was still concentrated in the city centers of the trading route on Java's north coast. Meanwhile, mountainous areas, especially in the interior of the southern part of Java, were still mostly uninhabited and took the form of forests.

The distribution of population towards Southeast Malang is inseparable from the development of Malang as *afdeeling* and gemeente. *Afdeeling* was one of the Dutch colonial government structure levels, which was equivalent to a district. Structurally, *Afdeeling's* position was part of the residency, whose leader was referred to as the assistant resident. One residency often consisted of several *afdeelings*. Meanwhile, gemeente was the name of an administrative division in a state structure equivalent to a municipality. *Afdeeling* Malang consisted of five controllers, namely Malang City, Batu, Kepanjen, Turen and Tumpang. Southeastern Malang was under the Turen district administration, which consisted of Turen, Wajak, Dampit, Ampelgading and Sumbermanjing. Administratively Afdeeling Malang was included in the Pasuruan Residency, which borders Surabaya, Kediri and Probolinggo.

3.2 Changes in landscape in Southeast Malang, 19th–20th Century

Southeastern Malang was administratively under Afdeeling Malang. The rapid development in this area occurred after the enactment of the Agrarian Law of 1870. A law born because of liberals in the Netherlands allowed private parties to play a role in land management in the Dutch East Indies and have erfpacht rights. Foreign investors could lease land for 75 years from *bumiputera,* and on average, the land was planted with industrial crops such as tea, coffee, sugarcane, cacao, quinine, cloves, indigo, and so forth. The enactment of this law had a major impact on the social and economic life in Java and Sumatra island. These impacts include monetization in Java, the growth of the plantation sector in South Bandung, the limited economic activity of Arabs in Surabaya (Utami 2015; Jayanto 2016; Hakim 2018). Meanwhile, in Malang, the enactment of this law impacted the mushrooming of private plantations, especially in Dampit and Ampelgading. Based on the Ampelgading map around the early 19th century, there were many cultivated plantations in this area such as *Og Soemberandong, Og Kaliglidik, Og Lebakroto, Og Soemberremis, Og Wonokojo, Og Soembersengkaring, Og Soembertelogo, Og Soembertjeleng, Og Sonosekar, OG Sonowangi,* and so forth.

The existence of this cultivation plantation (*onderneming*) has influenced the changes in the landscape and life patterns in Southeast Malang. There was a large-scale change of land use as a result, initially changing the wilderness into production plant land. In addition, the existence of these plantations invited the presence of people outside Southeastern Malang to settle there. Seeing Southeastern Malang's potential, the colonial government built transportation access to facilitate the transportation of agricultural products, among others, by making an axle road that could be traversed by cars from Malang to Ampelgading. In addition, a tram line was also built connecting Malang-Dampit as a continuation of the Lawang-Malang railway line, which was completed in 1879 (Handinoto 2004).

The conversion of land functions in the Southeast Malang region is an important marker for socio-economic changes in this region. The land is one of the natural resources that has significant benefits and functions in life because it is used for various things such as shelter and livelihood. Land often changes from time to time, according to how its used. Since the mid-19th century, there has been an increase in land use to cultivate various plantation products such as tea, coffee, sugarcane, quinine, cocoa, indigo and others. Accelerated due to European market demand, this phenomenon increased rapidly with the enactment of the Liberal Law, which allowed private parties to participate in the plantation industry. This activity changed the function of land in most of the islands of Java and Sumatra. In the *afdeeling* Malang, the visible changes are from forest land to plantation. Along with the rapid and almost total plantation coverage in Southeastern Malang, this certainly affected population growth. This population growth occurred due to migration from several areas outside *Afdeeling* Malang. This population displacement is caused by various things such as disease outbreaks, lack of employment opportunities, and the desire to improve living standards. Some of these reasons encourage population movement from one area to another. The migration that occurs in *Afdeeling* Malang was influenced by economic factors (Sayono et al. 2020). This can be seen from a large number of residents outside the area who have moved to work in the plantation sector as loggers to prepare the land, plant seeds, create plant nurseries and harvest plantations. Apart from plantation activities, the presence of these migrants was also influenced by the existence of plantation product processing factories in Southeast Malang and its surroundings, such as the Margosuko Coffee Factory, Wonokoyo Coffee Factory, Krebet Sugar Factory, Kebonagung Sugar Factory, et cetera.

Landscape changes in the Southeast Malang region are also affected by residential patterns. Prior to the 19th century, the settlement patterns in Southeast Malang were sporadically scattered around crowds such as markets and main roads. It was rare for the population to live near the highest slopes. However, the presence of plantations and processing industriesd caused settlement patterns to change. This change is especially evident in the Ampelgading area, where there are dozens of private *ondernemings* starting from Kaliglidik (the last village that is near the peak of Semeru) to Lebakroto, which is on the Malang-Lumajang axis. This settlement pattern is in the form of clusters of houses around big factories which are called *kampungan,* and special shelters in plantation complexes known as *persil*. The existence of these plantations and factories has changed the physical patterns of settlement and the composition of the population. Today, in Southeast Malang, we often find large families of the Madurese tribe who have lived for generations. They are mostly the fourth or fifth descendants of their ancestors who migrated to the region around the early 19th century. Based on the data collected by Gooszen about immigration in *Afdeeling* Malang, the largest population of migrants to this area are Madurese, then Javanese from various regions, especially Karesidenan Kediri and the Central Java region (Gooszen 1999). In addition, there are some Chinese ethnic migrants whose numbers can be counted on one hand. In Ampelgading, a family who are the fourth descendants of the Mbok Denok family can be found. They are Chinese people who traded in *Pasar* Jagalan, then married local residents and settled there. The number of ethnic Chinese people is more common in Dampit, especially around markets. Most of these ethnic groups work as collectors and distributors in the coffee trade.

3.3 *The effect of landscape changes on the life patterns of the Southeast Malang Community in the 19th–20th Century*

Changes in landscape and settlement patterns significantly affect the pattern of community life. This is due to the close relationship between humans and the environment. This environment's influence can be divided into two factors, namely the physical environment and the social environment. The physical environment here is the environment that covers an area's geography, while the social environment is the community environment consisting of the interactions between individuals (Walgito 1992). Additionally, changes in Indonesia's social environment after the Islamic sultanate were greatly influenced by the arrival of Europeans. This change began with trade interactions in the early 15th century and turned into colonialism, which lasted until the mid-20th century. Each region in Indonesia has different natural resources, ranging from spices to plantation crops.

Southeast Malang has never been affected by the glory of spices. This area became known by the colonial government for its soil fertility, which was suitable for plantation commodity crops around the early 19th century. Prior to the 19th century, it could be identified that the environment of Southeast Malang was mostly forest, which had an impact on the activities of the people who, on average, made a living by utilizing forest products, and it was possible that economically they still used the barter system in fulfilling their daily needs. Meanwhile, in the second half of the 19th century, when the big plantations were established, the people became working-class people who spent much of their time as plantation and factory workers. Various types of work emerged as a result of these changes in the environment, including *wiwil* and *gacok*. *Wiwil* is a coffee tree cleaner, who cleans bad coffee shoots, while *gacok* is a job as a craftsman that cleans plant roots around the coffee tree.

The new types of work in the plantation environment also accelerate social stratification based on professions in sequence, such as large foremen, area foremen, chant foremen and laborers. First-class, the big foreman, is the leader of several coffee plantation areas. Second class, the foreman *Isan* is a foreman who leads a plantation area. The third class, the *Rantek* foreman, is a small foreman who oversees several groups of workers according to their classification in one plantation area. The last class, the workers, are a group of workers specialized based on their daily work as cleaners, cultivators, carers and harvesters. This structured work is rewarded with a monetary salary. As previously explained, in almost all large plantation areas, the community recognizes monetization earlier than other communities. In Southeast Malang, workers' salaries were paid by means of a roller system by queuing for daily wages in front of *lodji*. In one day, female workers earned 1 *ece* (one level below the cent) and male workers attained 15 cents.

This new life pattern did not last long, only about 50 years. This change in the pattern was marked by the great depression of 1930. The crisis in America, which was preceded by the fall of Black Tuesday and the fall of the New York stock market on October 24, 1929, and reached its worst peak on October 29, 1929, had significant impacts on the world economy, including Southeast Asia (Rothemund 1996). In Southeast Asia, Java and East Sumatra were among the most affected because these two islands had large plantation industries owned by foreign investors under the Dutch colonial government's control (Siswoyo 2017). This phenomenon also occurred in Dampit and Ampelgading. Areas that were initially considered vital because they were cash crop producers had become neglected areas. This event was exacerbated by the arrival of Japan, which only focused on winning the Pacific War. Agricultural commodities whose sustainability was maintained included rice and corn to maintain food stability.

The absence of the maintenance of plantation crops in Southeast Malang caused many diseases, especially fungi that appeared on the roots, stems, and branches of crops. The events of independence and the revolutionary war resulted in resentful sentiments towards the remnants of colonialism, and great anger materialized in the form of burning plantation land, destroying facilities such as dams and water reservoirs and coffee processing factories. This further eliminated traces of the greatness of the plantation industry and cut the chain of knowledge in plantation cultivation. After the physical revolution, namely the Old Order and New Order periods, the Southeast Malang region became a region with no economic potential. This region has no uniqueness of industrial or agricultural products so that its development process is slow. In the New Order era, plantation lands left a perpetual agrarian conflict. The struggle for land use rights between Limited Liability Company Perkebunan Negara XII (PTPN XII) and the people resulted in neglecting these potential lands.

Slow changes in the ecological landscape in this area resulted in slow economic growth which led to people looking for new types of jobs that offered a better life, one of which was to travel and become a migrant worker. The people of the Southeast Malang region recognized this new profession since the early 1980s, and it was increasingly used as their main livelihood in the 1990s. Towards the end of the 20th century, the Southeast Malang region was known as the center of migrant workers. The consequence of this phenomenon is the waning popularity of agricultural work. Since then, this region's ecological landscape has not changed much because nature-based jobs such as agriculture and plantation work cannot directly bring wealth like being migrant workers in Saudi Arabia, Hong Kong, Singapore and Taiwan can.

4   CONCLUSION

Based on this research, the ecological changes that occured as a result of the economic development of The Take-off stage in Southeast Malang had a major impact on the pattern of its society's social and cultural lifestyle. The first ecological change was marked by the presence of industrial plantations under the domination of the colonial government, which resulted in changes of the forest landscape into industrial crops (coffee), the emergence of a new class (plantation workers) and monetization. Meanwhile, the second ecological change was marked by the destruction of coffee plantations as a result of the Great Depression and the Independence Revolution that resulted in a shift in social classes from the plantation worker communities becoming the transmigrant and migrant worker communities. This also turned out to bring a new culture like in the destination areas of migrants both domestically and abroad.

REFERENCES

Ghosh, D. K., Hossain, M. N., Sarker, M. N. I., & Islam, S. 2020. Effects of land-use changes pattern on tree plantation: Evidence from gher land in Bangladesh. *International Journal of Agricultural Policy and Research, 8*(June), 55–65.

Gooszen, H. 1999. *A Demographic History of the Indonesian Archipelago, 1880-1942.* Leiden: KITLV Press.

Hakim, C. 2018. *Politik Pintu Terbuka: Undang-Undang Agraria dan Perkebunan Teh di Daerah Bandung Selatan 1870–1929.* Ciamis: Vidya Mandiri.

Handinoto. 2004. Kebijakan Politik dan Ekonomi Pemerintah Kolonial Belanda Yang Berpengaruh Pada Morpologi (Bentuk Dan Struktur) Beberapa Kota Di Jawa. *Dimensi Teknik Arsitektur, 32*(01), 19–27. Retrieved from http://fportfolio.petra.ac.id/user_files/81-005/Kebijakan Politik.pdf

Hidayah, N., Dharmawan, A. H., & Barus, B. 2016. Ekspansi Perkebunan Kelapa Sawit Dan Perubahan Sosial Ekologi Pedesaan. *Sodality: Jurnal Sosiologi Pedesaan, 4*(3). https://doi.org/10.22500/sodality.v4i3.14434

Hudiyanto, R. 2015. Kopi Dan Gula: Perkebunan Di Kawasan Regentschap Malang, 1832-1942. *Jurnal Sejarah Dan Budaya, 9*(1), 96–115.

Imadudin, I. 2014. Dampak Kapitalisme Perkebunan terhadap Perubahan Kebudayaan Masyarakat di Kawasan Subang 1920-1930. *Patanjala, 6*(1), 65–80.

Iqbal, M., & Sumaryanto. 2016. Strategi Pengendalian Alih Fungsi Lahan Pertanian Bertumpu pada Partisipasi Masyarakat. *Analisis Kebijakan Pertanian, 5*(2), 167–182. https://doi.org/10.21082/akp.v5n2.2007.167-182

Jayanto, J. 2016. Industri Gula Di Karesidenan Cirebon Tahun 1870-1930 Dan Dampaknya Bagi Masyarakat. *Ilmu Sejarah – S1, 1*(1), 1–15. Retrieved from http://journal.student.uny.ac.id/ojs/index.php/ilmu-sejarah/article/view/4307

Kadir, H. A. 2018. Komparasi Munculnya Liberalisme Ekonomi di Indonesia dan Burma. *Lembaran Sejarah, 13*(2), 163. https://doi.org/10.22146/lembaran-sejarah.33541

Kuntowijoyo. 2005. *Pengantar Ilmu Sejarah.* Yogyakarta: Bentang.

Nigussie, Z., Tsunekawa, A., Haregeweyn, N., Tsubo, M., Adgo, E., Ayalew, Z., & Abele, S. 2021. The impacts of Acacia decurrens plantations on livelihoods in rural Ethiopia. *Land Use Policy, 100* (July 2020). https://doi.org/10.1016/j.landusepol.2020.104928

Petrenko, C., Paltseva, J., & Searle, S. 2016. Ecological impacts of palm oil expansion in Indonesia. *White Pater,* (July), 1–21. Retrieved from http://www.theicct.org/ecological-impacts-of-palm-oil-expansion-indonesia

Pirard, R., Petit, H., & Baral, H. 2017. Local impacts of industrial tree plantations: An empirical analysis in Indonesia across plantation types. *Land Use Policy, 60,* 242–253. https://doi.org/10.1016/j.landusepol.2016.10.038

Rostow, W. W. 1960. The Stages of Economic Growth. In *Political Studies* (Vol. 10). https://doi.org/10.1111/j.1467-9248.1962.tb00978.x

Rothemund, D. 1996. *The Global Impact of the Great Depression 1929–1939.* New York: Taylor and Francis.

Rustiadi, E. 2001. Alih Fungsi Lahan dalam Perspektif Lingkungan Perdesaan. *Lokakarya Penyusunann Kebijakan Dan Strategi Pengelolaan Lingkungan Kawasan Perdesaan, 10-11 Mei*(November).

Sayono, J., Ayundasari, L., Ridhoi, R., & Irawan, L. Y. 2020. Socio-economic impact in-out migration phenomenon in Southeastern Malang in 19th–20th. *IOP Conference Series: Earth and Environmental Science, 485*(1). https://doi.org/10.1088/1755-1315/485/1/012023

Siswoyo, T. 2017. *Pengaruh Malaise terhadap Perkebunan Kolonial Di Hindia Belanda Tahun 1930–1940.* Lampung: FKIP Universitas Lampung.

Utami, I. W. P. (2015). Monetisasi Dan Perubahan Sosial Ekonomi Masyarakat Jawa Abad XIX. *Jurnal Sejarah Dan Budaya, 9*(1), 51–63.

Walgito, B. 1992. *Pengantar Psikologi Umum.* Yogyakarta: Andi.

Watson, S. J., Luck, G. W., Spooner, P. G., & Watson, D. M. 2014. Land-use change: Incorporating the frequency, sequence, time span, and magnitude of changes into ecological research. *Frontiers in Ecology and the Environment, 12*(4), 241–249. https://doi.org/10.1890/130097

Community Empowerment through Research, Innovation and Open Access – Sayono et al (Eds)
© 2021 Copyright the Author(s), ISBN 978-1-032-03819-3

Reading culture development with optimizing digital library services during the pandemic

A. Asari*, I.A. Zakaria, A. Prasetyawan, M. Safii, T. Kurniawan, L.A. Rahmania & C. Fajar
Universitas Negeri Malang, Malang, Indonesia

ABSTRACT: ABSTRACT: This study aims to analyze the quality of the implementation of a digital library based on the Senayan library management system (SLiMS) application system to increase student reading interest in the MI Nurul Islam Bululawang Malang school library, starting from the quality of the system and constraints in the implementation process. The subject of this research is the librarian of the MI Nurul Islam school library, and the object of this research is the SLiMS application system. This research uses descriptive research with a qualitative approach. From the results of this study, it can be concluded that the MI Nurul Islam school library, in implementing digital libraries to support the development of student reading interest, has constraints on human resources, information and communication technology infrastructure in the implementation process of SLiMS-based digital libraries.

Keywords: reading, digital library, school library

1   INTRODUCTION

The digital library system is a library management process using information technology (IT) (Wenige & Ruhland 2018). The use of information technology in libraries aims to improve work efficiency and service quality to users (the right information, the right users, at the right time), related to the role and function of libraries as a force in the preservation and dissemination of scientific and cultural information that develops in line with human needs (Ahmad & Abawajy 2014). As the preservers of knowledge, the school library carries out activities which include hunting, collecting, identifying, managing, and disseminating information to the general public, which in its development can be assisted by communication and information technology equipment (Purwono 2008).

Libraries are institutions that manage collections of written works, printed works and/or recorded works professionally with a standard system to meet the needs of education, research, preservation, information and recreation for visitors (Basuki 2004). Library collections contain all information in the forms of written works, printed works and/or recorded works in various media forms that have educational value, which are collected, processed and served (Basuki 2020). Libraries have a function as a vehicle for education, research, conservation, information, and recreation to increase the intelligence and empowerment of the nation. The library aims to provide services to readers, foster a love of reading and broaden horizons and knowledge to educate the nation's people (Suwarno 2010). In Law No.43 of the year 2007 concerning libraries, article 12 paragraph (1) explains that library collections are selected, processed, stored, served, and developed by the interests of users by paying attention to developments in information and communication technology. Each library develops library services through advances in information and communication technology (Bafadal 2009).

Many libraries are constrained in carrying out the provisions of the above laws and regulations. Based on conditions in the field, the obstacles that arise are related to the infrastructure that supports

---

*Corresponding author: andi.asari.fs@um.ac.id

the library (Kusmayadi,2006). The availability of computers and library software is of course the main requirement in seeking the implementation of information technology in libraries (Jogiyanto 2007). Regarding the implementation of information technology in school libraries, in the present era, efforts have certainly been made (Kang & Jung 2018). However, for the implementation of automation systems and digital library systems, not all school libraries have implemented them (Gupta 2019). This is of course a common concern, to realize the implementation of information and communication technology in school library services (Pendit 2008).

However, not all libraries have implemented automation systems optimally. This is caused by several things. First, the constraints in determining the applications of a library automation system (Boyles & Meisinger 2020). Second, the ability to create personnel or human resources who have not received much training in the field of library automation systems (DeLone & McLean 2003). The library here is the MI Nurul Islam school library. Seeing the role of digital libraries in supporting the school library management process, it can be concluded that the SLiMS has an important role (Aswari et al. 2020). However, in practice, it is not necessarily a SLiMS that has the best quality and SLiMS does not always have a positive impact on the school library. Therefore, to find out the quality of the system SLiMS in the MI Nurul Islam school library needs to be examined and tested for the quality of its system (Petter 2008).

2   METHODS

This research used a qualitative approach. The data in this study was in the form of interviews with librarians and librarians in terms of system quality, information quality and service quality. The data sources that researchers used in this study were obtained from informants and documents. At this stage the researcher carried out the data collection process using the active participant method, namely the researcher collected data by being directly involved in the object to be studied. The data collection techniques used in this study were observation, interviews and document study. The data sources in this study were obtained based on information about the implementation of the digital library in the MI Nurul Islam school library. The main data sources are library students, teachers and librarians for a total of eight people from a total of 50 visitors. These eight people were chosen because they have accessed the digital library system (Mukhtar 2013).

3   RESULTS AND DISCUSSION

3.1   *Quality of library automation systems in support of reading culture*

In this study, the quality of the SLiMS is measured based on criteria according to Hamilton in Jogiyanto (2007). There are eight indicators, (a) a proposed data currency: in this study, the capabilities of the SLiMS information system in displaying library data or information can always be updated properly; (b) response time: in this study the SLiMS was used to process work performed by system users, in this case, the library work done by librarians was fast enough. The SLiMS has a fast response time in processing commands from system users or system operators; (c) replacement time is the speed of the SLiMS, in this study the SLiMS has a fairly fast feature change time, such as changing menus that have a fairly fast processes; (d) data accuracy, in this study the accuracy of the information or data displayed by the SLiMS is determined by the system user. The SLiMS has a fairly high level of data accuracy in displaying data or information; (e) reliability, in this study, it was found that the SLiMS can be relied upon in the process of completing tasks or work in MI school libraries Nurul Islam. SLiMS can be used in library management from collection management to collection services; (f) completeness, in this study it was found that the SLiMS has quite complete features in supporting the management of the MI Nurul school library and the completeness of the menu features that exist in the SLiMS strongly supports the management and service processes for users of the MI Nurul Islam school library; (g) the flexibility: flexibility means the SLiMS can be applied to the MI Nurul Islam library and its unique conditions, such as the condition that the library has a limited number of librarians, so using the SLiMS can take some pressure off library staff. In terms of system access, the SLiMS can be accessed using ethernet and internet-based computer networks. Finally, (h) ease of use: in terms of ease of use the SLiMS is quite easy to understand. In this study, it can be concluded that the operation of the SLiMS is quite easy and it can be used by librarians and librarians without training.

3.2 *Quality of information to support reading culture*

The quality of the information is part of the indicators to be measured. The quality of the information in question is the quality of information issued or produced by the SLiMS. In this study, the quality of information is measured by Mukhtar's (1999) opinion that information must be reliable, relevant, timely, complete and understandable. A complete explanation is as follows. It must be, (a) reliable: from the research results it can be found that the SLiMS in the MI Nurul Islam library has timely quality information when needed by the user. So the information released by the SLiMS is reliable. It must be, (b) relevant: in the process of searching for information on the SLiMS, it must displayed information that is relevant to users. From the results of the study, it can be concluded that the SLiMS always displays the information desired by system users, and the information released by the SLiMS is suitable and appropriate to user requests in the MI Nurul Islam library. It must be, (c) timely: the SLiMS is punctual and accurate. From the research results, it can be concluded that the SLiMS can present information on time according to the needs of system users. It must be, (d) complete: the completeness of the information displayed by the SLiMS is quite complete. From the research results, it can be concluded that the SLiMS, when displaying information, is complete according to the user's wishes. It must be, (e) understandable: in terms of convenience, it can be concluded that the SLiMS has information that is easy to understand by users.

3.3 *Quality of service in supporting a reading culture*

The quality of service in this case is the quality of service that can be provided by the SLiMS to system users. This study refers to the opinion of Jogiyanto (2007). That there are five indicators used to measure the quality of services in the SLiMS. These indicators include, (1) reliability: reliability in this case is the user's perception of the reliability of the SLiMS. From the research results, it can be concluded that the SLiMS can be relied upon in completing the work and tasks of managing the MI Nurul Islam library. The next indicator is (2) responsiveness, responsiveness in this case is the user's perception of the responsiveness obtained by the user in completing tasks using the SLiMS. From the results of the study, it can be concluded that librarians using the SLiMS have a high enough responsiveness to immediately complete work by utilizing services in the SLiMS. (3) is assurance, in this case, the guarantee is the user's perception of ease when using the SLiMS. From the research results, it was found that librarians using the SLiMS can be helped and the librarian's work becomes easier to complete. (4) is empathy, empathy here is the system user's perception of the suitability of features or menus in the SLiMS. From the research results, it can be concluded that the features or menus in the SLiMS are quite complete following the expectations and desires of system users. The next indicator is (5) tangibles, in this case, tangibles are the physical components of the implementation of the SLiMS in the MI Nurul Islam library. Physical components here includes two aspects, namely computer equipment and computer networks. From the research results, it was found that the implementation of the SLiMS in the MI Nurul Islam library was supported by computers that were of moderate quality, and the number was still minimal, namely, there was only one computing device. Then, the network quality used by the MI Nurul Islam library is classified as not good because it still uses a wifi network that has a small capacity so that the speed is slow. This can be a problem in the process of implementing the SLiMS which has the effect of reducing the quality of MI Nurul Islam library services. Therefore, these two aspects need to be improved both in terms of quality and quantity, to support the performance and good service of the MI Nurul Islam library.

4 CONCLUSION

Based on the results of research that has been done, it was found that the implementation and development of the SLiMS in the MI Nurul Islam library in terms of system quality is quite good because it can always display updated information so that it can foster reading interest, it has a fairly fast response time and feature change time, the information displayed is always accurate, has complete features, is easy to operate and can be accessed using ethernet and internet networks, making it easier for users to get information and have an impact on increasing user reading interest. Then, in terms of information quality, the SLiMS can provide accurate and complete information. However, the quality of the content shown by the SLiMS is still sometimes poor. Furthermore, in terms of service quality, the SLiMS can make work easier and more reliable in the MI Nurul Islam library; the features and menus in the SLiMS are helpful in supporting library management. The use of the SLiMS in the MI Nurul Islam library can make librarians more responsive in the process of completing work in the library. However, the implementation of the SLiMS in the MI Nurul Islam library is not supported by a good computer infrastructure and internet network. There is only one computer so it does not support user service processes and the existing internet is of poor quality.

REFERENCES

Ahmad, M., &Abawajy, J. H. 2014. Digital library service quality assessment model. *Procedia-social and behavioral sciences*, 129: 571–580.

Aswari, M., Kristiawan, M., &Fitria, H. 2020. Senayan Library Management System Application as Digital Library Management. *International Journal of Progressive Sciences and Technologies*, 20(2): 129–136.

Bafadal, Ibrahim, 2009, *School Library Management*. Jakarta: Earth Literacy.

Basuki, S. 2004. *Perpustakaan Nasional dan Asosiasi Pustakawan di Indonesia dilihat dari Segi Sejarah.* The National Library and the Indonesian Librarian Association: a historical overview, in Temu Ilmiah Berdirinya Perpustakaan Nasional RI dan Peran Organisasi Profesi.

Basuki, S. 2020. Pengertian Pustakawan Menurut Perundang-Undangan Indonesia Serta Berbagai Dampaknya. *Media Pustakawan*, 16(2): 52–57.

Boyles, J. L., &Meisinger, J. 2020. Automation and Adaptation: Reshaping journalistic labor in the newsroom library. *Convergence*, 26(1): 178–192.

Delone, W. H., & McLean, E. R. 2003. The DeLone and McLean model of information systems success: a ten-year update. *Journal of management information systems*, 19(4): 9–30.

Gupta, M. K. 2019. Library Automation: Issues, Challenges and Remedies. *International Journal of Techno-Management Research*: 19–25.

Jogiyanto, 2007. *Success Models of Information Technology Systems*. Yogyakarta: Andi

Kang, B. S., & Jung, Y. 2018. Awareness on the Establishing and Operation of the Makerspaces in School Libraries. *Journal of the Korean Society for Library and Information Science*, 52(3): 171–192.

Kusmayadi, Eka. 2006. Online Study Public Access Catalog (OPAC) in Library Services and Agricultural Technology Dissemination. *Agricultural Library Journal*, 15(2).

McGill, T., Hobbs, V., & Klobas, J. 2003. User developed applications and information systems success: A test of DeLone and McLean's model. *Information Resources Management Journal (IRMJ)*, 16(1): 24–45.

Mukhtar. 2013. *Practical Methods of Qualitative Descriptive Research*. Jakarta: Reference (GP Prees Group)

Pendit, P., 2008. *Digital Libraries from A - Z*. Jakarta

National Library of Indonesia. 2014. *INLIS Lite User Manual Version 2.1.2 (Integrateg Library System)*. Jakarta.

Petter, Stacie., William H DeLone and Ephraim R. McLean. 2008. Measuring information systems success: models, dimensions, measures, and interrelationships. *European Journal of Information Systems*. 17: 236–263.

Salmon, Stephen R. 1985. *Library Automation Systems, Marcel Dekker*, New York

Suwarno, Wiji. 2010. *Library Science & Librarian Code of Ethics*. Yogyakarta: Ar-Ruzz Media.

Wenige, L., & Ruhland, J. 2018. Retrieval by recommendation: using LOD technologies to improve digital library search. *International Journal on Digital Libraries*, 19(2–3), 253–269.

Community Empowerment through Research, Innovation and Open Access – Sayono et al (Eds)
© *2021 Copyright the Author(s), ISBN 978-1-032-03819-3*

Those who are forgotten: The existence of *Ketoprak Rukun Karya* in Sumenep Madura, 1976–2000s

J. Sayono*, R. Ridhoi, N. Jauhari, I.H. Al Siddiq, A. Prasetyawan & N.A.D. Restanti
Universitas Negeri Malang, Malang, Indonesia

ABSTRACT: Since the mid-20th century, the traditional art of *ketoprak* on Madura Island, particularly in Sumenep Regency, has shown its existence. There are two well-known *ketoprak* in Madura, namely Rukun Famili and Rukun Karya. However, the original Sumenep *ketoprak* art seems absent from Indonesian and Maduran historiography. This paper aims to explain the history of traditional theater arts in Madura, especially the Rukun Karya *ketoprak* located in Sumenep. This paper uses the historical method by conducting a careful reading of several archival sources and interviews. The results of this study indicate that although *Ketoprak* Rukun Karya had existed since 1976, the people of Sumenep (urban) do not know and like this art. Additionally, in the Sumenep area, the Rukun Karya archipelago has become a popular and luxurious entertainment for the people. This is due to the rapid modernization of urban society and the erosion of local artistic values.

Keywords: *Ketoprak*, Rukun Karya, Sumenep, Madura Island

1   INTRODUCTION

Indonesia's wealth lies not only in its natural and human resources but also in its cultural diversity, which must be preserved. Each Indonesian region has a variety of arts. These arts are usually passed down from generation to generation so that they can still be enjoyed by the younger generation (Weintraub 2010; Ridhoi 2018; Azali 2012). In relation to the traditional theater arts, Sumenep Regency is one of Madura's areas that still has one of the traditional theater arts, namely *ketoprak*. The local community is more familiar with this *ketoprak* as *loddrok*. Since around the middle of the 20th century, several *ketoprak* have been born on Madura's island to be precise in the area of Sumenep Regency.

The famous *ketoprak* in Madura is the *ketoprak* Rukun Family and Rukun Karya. Both *ketoprak* have been around for a long time and are still in existence today. Similar to the general *ludruk* and *ketoprak*, the *ketoprak* in Madura also take stories from everyday life and from history and legends. *Ketoprak* Madura also has a director or puppeteer to arrange the storyline during the show (Imron 2020). This article aims to investigate the development of traditional theater arts in modern times like today.

Several previous studies discuss traditional theater arts in East Java. One of them is written by Samidi, entitled Traditional Theater in Surabaya 1950–1965: Community Relations and Art Troops. This article describes the development of traditional theater in the mid-20th century in Surabaya. Traditional theater performances in Surabaya are closely related to the relationship between the utterance of solidarity with the community (Samidi 2006). The article's goal is to find out the relationship between the community or the audience and a group of traditional theater players in Surabaya in the mid-20th century.

*Corresponding author: joko.sayono.fis@um.ac.id

In addition, there are also articles that discuss the development of *ketoprak*; one article is written by Saptomo, entitled History, and Development of *Ketoprak* in Modern Society Life. The article describes the development of *ketoprak,* which has changed according to the conditions of the community (Cohen 2013). According to Saptomo (1996), the community's development and attention to the art of *ketoprak* theater was only felt in the 1950s. From those several previous studies, there are very few researches that focus on the history and development of the Rukun Karya *ketoprak* in Sumenep. The history of traditional theater arts must be investigated deeply. This is because the development of the conventional theater arts era needs preservation to be enjoyed by young people and gain more attention from the Indonesian people.

2   METHODS

In writing this article, researchers used historical research methods by utilizing sources such as interviews, articles, and so forth. The historical research method has five stages, namely topic selection, source collection (heuristics), source verification, interpretation and writing. The topic chosen by the researcher was a research plan that had been prepared beforehand, namely by selecting the topic of art history, especially the art of *ketoprak* in Sumenep Regency, Madura, East Java. After determining the topic, the researcher collected sources by collecting news about *ketoprak* performances in Madura, books that discussed *ketoprak* or traditional theater arts and interviews with *ketoprak* figures in Rukun Karya in Sumenep. After that, source verification was completed by looking at the credibility of the sources obtained. The sources were classified into two categories, namely primary sources or secondary sources. After verifying the source, the interpretation was carried out by analyzing the obtained sources and continued to the source criticism stage. Finally, a writing or historiography process regarding the history of the traditional theater was conducted. It discussed the development of the Rukun Karya *ketoprak* in Sumenep in the 2000s.

3   RESULTS AND DISCUSSION

3.1   *Development of traditional theater arts*

Theater is an art that has long been developing in Indonesia. It is divided into three types widely known by Indonesian people: traditional theater, transitional theater and modern theater (Peters 2016). The traditional theater is a theater that was born from, by, and for the traditions of a particular society (Heryanto 1982). The traditional theater in each region is different. The term traditional theater is part of a regional art form that was born and developed in certain communities and those who live in certain areas (Zaini 2015). One example is the traditional theater in East Java, which is better known as *ludruk* by the community.

Besides *ludruk*, there is also a traditional theater art that has developed in Indonesian society, *ketoprak*. The famous *Ketoprak* originates from the Central Java area. But along with the development of the era, *ketoprak* also develops anywhere in all regions of Indonesia. According to the research results of the Art Section of the Cultural Service (Art Division of the Directorate General of Culture), *ketoprak* was created in Central Java, more precisely in Surakarta, in 1908 by RMT Wreksodiningrat. The name *ketoprak* also comes from the name of a *tiprak* musical instrument (a type of bamboo musical instrument used to repel birds in the fields) (Saptomo 1996).

Ketoprak has a long history. Herry Lisbijanto describes the periods of *ketoprak* as it evolved. The first is the *ketoprak* Gejog or Lesung period in 1887–1908. Initially, this art was a type of play by village youths to joke around during the full moon by singing traditional Javanese songs or *dolanan* songs and was accompanied by sounds that came from *lesung*. Second was the Wreksadiningrat *ketoprak* period from 1908–1925. A royal servant of the Surakarta Hadiningrat palace named KRMT H Wreksadiningrat worked on this *ketoprak* art to turn it into the pride of palace art. The third is the Wreksatama *ketoprak* period from 1925–1927 when a new ketoprak group was established outside the palace standing in Madyataman Surakarta. It was founded by Ki Wisangkara (a former member of the Wreksadiningrat *ketoprak* group). The fourth is the Krida Madya Utama period of 1927–1930 which was a ketoprak group founded on the wishes of the community itself. The fifth is the Gardanela *ketoprak* period from 1930 to 1955 which was the ketoprak period that experienced many novelties in the form of stories, costumes, and so forth. The sixth is the modern *ketoprak* period from 1955–1958. This *ketoprak* period was often referred to as the *ketoprak* tobong, because it often changed the performance location. The last one was the *Ketoprak* Gaya Baru period in 1958–1987. There were many innovations from *ketoprak* artists in this period to attract the public's interest to watch *ketoprak* performances (Lisbijanto 2013).

Based on the description above, which briefly explains the traditional theater arts, namely *ludruk* and *ketoprak*, the two traditional theater types have almost no differences. One of the examples lies in the stories that are being told, which both taking stories from everyday life (Saptomo 1996). However, there is a little difference between the play's meaning in the performance of ludruk and *ketoprak*. Another supporting aspect is the use of language during performances. *Ludruk* and *ketoprak* often use colloquial language or languages from the area of origin, for example, using Javanese, Madurese, or other languages.

3.2   *Ketoprak Rukun Karya and its development in Sumenep*

Sumenep is an area that has several cultures and arts that must be preserved. Some of the arts in the Sumenep Regency include building art, sports, dance, drama and puppetry (Prawirodiningrat 1986). Mr. Tajul Arifin (2020), who is a cultural observer and the Head of the Cultural Heritage of Sumnep Regency, Madura, East Java, also explains that there is quite a lot of culture and art in the Sumenep Regency. One of them is culture and also art from the heritage of the Sumenep Palace. *Ketoprak* is a traditional theater and drama or puppetry that continues to develop in Sumenep Regency, Madura, East Java. The Madurese community is more familiar with the Madura *ketoprak* as *loddrok*. Before being known as *loddrok*, this performance was known as *ajhing* (Santoso 2016). As supported by Mr. Tajul Arifin (2020), before being called loddrok, this art's name was ajhing. Ajhing is one of the performances that bring a prayer of goodness played by a group of men and accompanied by a *saronen* orchestra, which is staged from village to village and is followed by jokes about daily life taken from the story of One Thousand and One Nights (Bouvier 2002).

Two ketoprak groups that are still active in Madura are the Rukun Famili and the Rukun Karya *ketoprak* groups. *Ketoprak* in Madura has existed since Dutch rule era. However, at that time, it was still known as ajhing. In relation to Madura's traditional theater arts, there are two terms that are often used, namely *ketoprak* and *loddrok* or *ludruk* (Sayono, et al. 2020). The Madurese community and some writers often use the term *loddrok* or *ludruk* instead of using the term *ketoprak* to refer to these two traditional theater groups in Madura. This is because, according to research, the term "*ketoprak*" is not included in the Madurese language, unlike *ludruk,* which later became *loddrok* (Bouvier 2002). Then around the 1970s *loddrok ketoprak* appeared during Yogyakarta's official seminar on *ketoprak* in 1977.

According to Mr. Encung Hariadi (2020), as the director of the Rukun Karya *ketoprak*, it has been in Sumenep since 1949. Three ketopraks already existed at that time, including the *ketoprak* Banjir Dunia, Bintang Massa and Rukun Muda (Hariadi 2020). In the 1960s, a number of *loddrok* had also existed, namely Jata Kemala (from the village of Juruan Daya in Batuputih), Karya Kemala (from Juruan Laok) and then *Loddrok* Karya Putra was born (Bouvier 2002). Some of these *loddrok* groups still play ajhing in their performances. According to Bouvier (2002), the Madurese *loddrok* was inspired by elements of old ajhing jokes: word play, mimicing, bodybuilding and black and white makeup. The difference between *ketoprak* and *loddrok*, according to the *loddrok* group from Batuputih, can be seen in the clothes used, such as the king's *loddrok* who wears a long-sleeved shirt while in *ketoprak*, they are bare-chested and only wear jewelry (Bouvier 2002).

Since the time of Dutch rule, there have been *ketoprak*s in Madura. One of them is the Rukun Santoso group from Tanjung Village, Sumenep Regency, East Java. At that time, the performances were still straightforward, done without decorations, only playing for the village people and in the

Figure 1. *Ketoprak Rukun Karya* performance.

form of a play, not *ketoprak* (Bouvier 2002). Close to the independence era, around 1943, the name Rukun Santoso was changed to Rukun Famili and switched from theatrical to *ketoprak* (Bouvier 2002). The Rukun Famili group also uses accompanying music with gamelan instruments. It can be said that the Rukun Famili *ketoprak* group marked the beginning of the birth of other *ketoprak* groups in Sumenep Regency, Madura, East Java. In 1975, a member of the Rukun Famili group went to do an internship at the Siswo Budoyo *ketoprak* group for one month. At the end of 1975, the Rukun Famili *ketoprak* group was split into two groups, namely the Rukun Famili and the Rukun Karya, and applied the knowledge that had been acquired during an internship at Siswo Budoyo in 1977 (Bouvier 2002).

Ketoprak Rukun Karya was born in 1975 and began to develop around 1976 as a fraction of the Rukun Famili *ketoprak* group. According to Mr. Encung Hariadi (2020), as the director of the Rukun Karya *ketoprak* group, the Rukun Karya *ketoprak* was born in 1976, while the name Rukun Karya was taken because at that time Golkar Party was popular (*Golongan Karya*). Based on the results of an interview with Mr. Encung Hariadi (known as Mr. Jaka Linglung), who is currently serving as the director of *ketoprak* Rukun Karya, at the beginning of the show, to promote the *ketoprak*, they still used tickets to watch the *ketoprak* performance. Currently, the *Ketoprak* Siswo Budoyo from Tulungaggung has joined the *Ketoprak* Rukun Karya for about a month. Its director since 1994, Mr. Encung, explains that a director in *ketoprak* Madura has to be really positive because the audience assessed and researched what is delivered in the *ketoprak* show. In the early days of the Madura *ketoprak*, which was around the 20th century, the narrative model during the performance had to be truly pure in presenting a story, be it historical stories, legends or everyday life. Then around the 21st century, from 2007 to 2008, during the Rukun Karya *ketoprak* show, they slipped in a little joke before telling the story. Similar to the stories presented in general *ketoprak*'s, the Rukun Karya *ketoprak* also brings historical stories such as the history of Sumenep or Pamekasan, Majapahit, Walisongo, and others.

Figure 1 shows a sequence of a comedy scene from Rukun Famili *ketoprak* performances in 2020. During this Rukun Karya *ketoprak* performance, the local language of Maduranese is often used. The use of regional languages is also a special characteristic of *ketoprak* or *ludruk* performances in various regions. In Madura the *ketoprak* show can also be staged in various conditions. For example, it is held in public places using tickets, for private ceremonies such as weddings, and for spiritual events or at sacred graves (Bouvier 2002). In performances at weddings, the Rukun Karya *ketoprak* starts at 23.00 and runs to 03.00 in the morning. Before that, Javanese repertoire in the form of a set of *gamelan* was already heard.

As is well known, several traditional arts, including theater, music, dance, and others, have declined due to changing times. It cannot be denied that the decline in public interest in watching traditional theater was one of the triggers for the collapse of the art theater (Spiller 2012). Therefore, the Rukun Karya *ketoprak* originating from Saronggi, Sumenep Regency, Madura, East Java is struggling to maintain its existence. There are several things that make the Rukun Karya *ketoprak* group survive, such as their name that has been well known among the Madurese community even in other areas such as in East Java and West Java, their form of an organization, and the existence of AD/ART (Articles of Association/Budget Household) (Hariadi 2020).

3.3 *The development of traditional theater arts in the modern era*

At present, the types of theater in Indonesia are increasingly developing following the changing times. Some traditional theaters are no longer visited by Indonesians—consequently, several other theaters were born, known as modern theaters in Indonesia. The modern theater has different characteristics from traditional theater. In modern theater the presented stories prioritize western theater aesthetics or dramaturgy, where performances are held in a special place (the proscenium stage) and the audience has to pay an entrance ticket. While it functions only as entertainment, the story is told more on the topic of contemporary events, and the usual language used is Indonesian (Zaini 2015). However, at the present time, traditional theater also tries its best to survive. Therefore, several modern theater groups have remodeled some of their performance elements to make them more interesting and less boring. One example is traditional theater, such as *ketoprak*.

Currently, the lighting on the *ketoprak* show is awe-inspiring. In fact, due to good lighting, the *ketoprak* show is no longer perceived as traditional theater but has become a modern theater. Not only that, other traditional theaters such as *ketoprak*, *ludruk*, and wayang have also begun to be packaged in cassette format, broadcast on radio and television. This shows the traditional theaters' persistence in maintaining their existence by keeping up with the changing times. Even though there have been developments in terms of performances, the contents of stories presented by traditional theater such as *ketoprak* still use royal stories, legends and still contain moral messages.

In general, *ketoprak* has also undergone changes and developments from time to time. First, it can be seen from the musical instruments used to accompany the *ketoprak* show. Initially, the *ketoprak* accompaniment music used a mortar, but now it uses Javanese gamelan. Second, it can be seen from the stories that are delivered, the beginning of the establishment of *ketoprak* only brings stories of everyday life until later it also delivers royal stories, histories, legends and so forth. Another prominent change is observed in the development of *ketoprak* costumes. At first, the costumes used were only everyday clothes, but when the *ketoprak* had developed royal stories in their performances, the costumes used were adjusted to the ones that were to be played. In addition, the stage for the performance has also changed; the backdrop used has been filled with lighting and depicts an atmosphere such as a palace, forest, garden, and so forth (Lisbijanto 2013). It can be seen in Figure 1 that illustrates the performance of Rukun Karya *ketoprak* in 2020. In the photo, the backdrop used is beautiful and matches the background of the presented story.

In the 1980s–1990s the art of *ketoprak* grew with new innovations. During those years, a *ketoprak*, which was commonly called *ketoprak plesetan* existed (Lisbijanto 2013). In this type of *ketoprak*, performance is packaged more loosely, and the characters played more varied. Besides that, the addition of humor to the show is also a development in *ketoprak* (Hatley 1971; Peacock 1967). In 1995, the art of ketoprak also grew, which showed its comedic function, namely the *ketoprak* humor tradition (Lisbijanto 2013). This *Ketoprak* is usually performed on national television with not many *ketoprak* players. Some of the innovations that grew in the 1990s-2000s has helped *ketoprak* to survive (Sunardi 2011)

4 CONCLUSION

Along with the development of the era, traditional theater almost disappeared in Indonesian society. Public engagement is an essential factor in the survival of conventional theater. The decreasing public interest in watching or maintaining the art will indirectly erase traditional theater. Modern theaters have developed a lot in Indonesia. Therefore, traditional theater groups carry out various reformations or innovations to survive in the modern era, like *ketoprak,* which currently adds jokes before the story begins. Not only that, but the stage part has also been arranged in a more modern way in order to captivate the *ketoprak* audience. The addition of lighting and stage backdrops has also been adjusted to the story's background to attract the audience's attention. Therefore, apart from the need for innovations from traditional theater artists, the public is also an essential factor that can maintain the existence of traditional theaters in Indonesia.

REFERENCES

Azali, K. 2012. *Ludruk*: Masihkah Ritus Modernisasi?. Lakon: Jurnal Kajian Sastra dan Budaya, 1(1).

Bouvier, H. 2002. Lebur: Seni Musik dan Pertunjukan dalam Masyarakat Madura. Bogor: Mardi Yuana Graphic Printing.

Cohen, M. 2013. Anthologizing Indonesian Popular Theatre. *Asian Theatre Journal,* 30(2), 506–519. Retrieved December 1, 2020, from http://www.jstor.org/stable/43187275

Hatley, B. 1971. Wayang and *Ludruk*: Polarities in Java. The Drama Review: TDR, 15(2), 88-101. doi:10.2307/1144625

Heryanto, A. 1982. Teater di Indonesia. Basis, November.

Lisbijanto, H. 2013. *Ketoprak*. Yogyakarta: Graha Science.

Peacock, J. 1967. Comedy and Centralization in Java: The *Ludruk* Plays. *The Journal of American Folklore,* 80(318), 345–356. doi:10.2307/537413

Peters, R. 2016. Death and the Control of Life in an Indonesian City. Bijdragen Tot De Taal-, Land- En Volkenkunde, 172(2/3), 310–342. Retrieved December 1, 2020, from http://www.jstor.org/stable/44325706

Prawirodiningrat, SI. 1986. *Sepintas Kilas Adat Budaya Sumenep Sebagai Aspek Pembangunan Nyata.* Sumenep: Solar Offset Printing.

Rahayu, F. 2014. Perkembangan Seni Pertunjukan *Ludruk* di Surabaya Tahun 1980-1995 (Tinjuan Historis Grup Kartolo CS). *Avatara*, Volume 2, No. 2.

Ridhoi, R. 2018. Melihat Motif Pendidikan Di Nusantara Dari Perspektif Historis. *Jurnal Pendidikan Sejarah Indonesia*, 1(2), 135-149. Retrieved from http://www.jurnalpsi.com/index.php/jpsi/article/view/15

Samidi. 2006. Teater Tradisional di Surabaya 1950-1965: Relasi Masyarakat dan Rombongan Seni. *Humanities,* vol. 18, No. 3.

Santoso, M. 2016.Transformasi Bentuk Tari Srimpi dalam Pembukaan *Loddrok* Rukun Famili di kabupaten Sumenep-Madura. From (Jurnalmahasiswa.unesa.ac.id, accessed on 06 September 2020).

Saptomo. 1996. Sejarah dan Perkembangan *Ketoprak* dalam Kehidupan Masyarakat Modern. Education Horizons, Number 2.

Sayono, J, et al. 2020. *Dari Ajhing Hingga Ketoprak: Perjalanan Historis Kesenian Ludruk di Sumenep Madura Sejak 1940-an*. Malang: Java Creative.

Spiller, H. 2012. How Not to Act like a Woman: Gender Ideology and Humor in West Java, Indonesia. *Asian Theatre Journal,* 29(1), 31–53. Retrieved December 1, 2020, from http://www.jstor.org/stable/23359543

Sunardi, C. 2011. Negotiating Authority and Articulating Gender: Performer Interaction in Malang, East Java. *Ethnomusicology*, 55(1), 32–54. doi:10.5406/ethnomusicology.55.1.0031

Weintraub, A.N., 2010. *Dangdut stories: a social and musical history of Indonesia's most popular music.* Oxford University Press.

Zaini, M. 2015. *Pembelajaran Seni dan Teater untuk Siswa. College Students, and the Public*. Yogyakarta: Frame Publishing.

Encung Hariadi (38 y.o.), interviewed on August 29, 2020.

Tajul Arifin (58 y.o.), interviewed on 31 August 2020.

Zawawi Imron (75 y.o.), interviewed on July 18, 2020.

Community Empowerment through Research, Innovation and Open Access – Sayono et al (Eds)
© 2021 Copyright the Author(s), ISBN 978-1-032-03819-3

Religious discourse, cyberspace and social media: A trajectory from Muslim millennial's perspective

T. Thoriquttyas* & N. Muyassaroh
Universitas Negeri Malang, Malang, Indonesia

N. Ahsin & A. Naim
IAIN Kediri, Kediri, Indonesia

ABSTRACT: The various interpretations of the Quranic verses on social media is still a contextual topic to be discussed. In fact, Muslim millennial groups often rely on social media as a way of learning Islam (i.e. Twitter, Facebook, Instagram etc). The purpose of this study is to snapshoot the trajectory of religious discourse in cyberspace from a Muslim millennial perspective. In this study, the Muslim millennial group was represented by the freshmen of Universitas Negeri Malang (UM). Data was collected through interviews and in-depth observation of 145 students who studied Islamic education's course from Faculty of Letters, Faculty of Sports Science and Faculty of Social Sciences. The results obtained show that there is a relationship between respondents' responses on religious discourse and social media. The various responses are indicating that Muslim have varied views relating to the intersection of religious discourse and social media

Keywords: cyberspace, social media, religious discourse

1   INTRODUCTION

Social media has the potential to become a primary source for accessing religious knowledge and disseminating moderatism as well as radicalism to the Muslim millennial generation. Again, discussions about the relationship between the internet and religion have great significance. Discussing moderatism and radicalism in religious views from cyberspace is a complicated topic (Baulch & Pramiyanti 2018; Hussain & Silcock 2019). It needs a multidisciplinary approach to uncover the dynamic of religious discourse in cyberspace. In the Indonesian context, the prevalence of religious discourse on social media is wide-spread, more so for Indonesian Muslims (Faiqah & Pransiska 2018).

As predicted by Hootsuite (*We are Social*) in 2017, the users of social media in Indonesia reached 160 million users in January of 2020. The number of social media users in Indonesia increased by 12 million (8.1 percent) between April of 2019 and January of 2020 (Slama 2017). Social media use in Indonesia was at 59 percent in January of 2020 (Herdiansah & Husin 2018; Slama 2018). In the Indonesian context, there are the five popular social media platforms in 2020: YouTube (88%), WhatsApp (84%), Facebook (82%), Instagram (79%) and Twitter (56%). With this number of users, the Muslim millennial generation becomes the majority group who is involved in social media.

This project portrays the social media use of Indonesian Muslims on the discourse of moderatism and radicalism. This warranted an approach to research that enabled young people to discuss their views and experiences, explain why they engage in moderatism or radicalism and give their motivations for doing so.

---

*Corresponding author: titisthoriq.fs@um.ac.id

DOI 10.1201/9781003189206-21

Ahmad Syafii Maarif stated that radicalism cannot be defined as terrorism (Suhaimi & Raudhonah 2020). According to him, radicalism is more concerned with the way religious person acts and the accumulation of model attitudes daily (Arifianto 2019; Irham et al. 2020). On the other hand, the term of terrorism includes criminal acts and has political objectives. In addition, the phenomenon of radicalism is more related to religious cases and terrorism requires global action. Radicalism can sometimes transform into terrorism when the circumstances allow (Woodward 2010).

The formulation of radicalism always adjusts its linearity to the trajectory of the times (Azra 2005; Nur 2020). Several years ago, acts of radicalism in the name of religion often used physical force as a tool (Nur 2020). However, this is currently changing. The current portrait of radicalism is often narrated in social media using virtual instruments. The virtual instrument can be interpreted as the model of distributing and disseminating religious content through social media.

It is interesting to note that there are new modifications in the spread of radicalism. One of them is the emergence of radical Islamic books in the market. These books have unique characteristics especially when it comes to the type of publisher. The book's publisher has ideological closeness to the author or the book's themes, so that according to Azra, there is a close relationship between the publisher and the book's themes that have radical content (Azra 2005; Fealy 2012; Salem 2016). This da'wah method is used specifically for conservative Islamic groups. According to the Salem, the kind of publications from conservative Islamic circles can be easily found and accessed by the public, such as magazines, pamphlets, VCDs and any content uploaded on their website (Salem 2016).

Winarni in her research shows that the radicalism that is currently developing attracts religion, especially Islam in situations and conditions that are inevitable and as if it creates connections between the interpretation of Islam and violence (Winarni et al. 2017; Winarni et al. 2019). This is certainly detrimental to Islam and all its components, because it portrays Islam as tough, stiff and "angry" (Winarni 2014). The thing that must be remembered and pondered deeply is that the birth of Islam in its long historical trajectory was never colored with blood and decorated with sharp swords, but Islam was brought by the Prophet Muhammad with messages of peace.

This study examines social media as it currently takes a very large portion and role in providing good information that is directed at radicalism and moderatism, especially with segmentation for the millennial generation. Both moderatism and radicalism also use social media lines to disseminate their thoughts. This is emphasized by the enthusiasm of the millennial generation, especially when it comes to digital literacy. Propaganda and the recruitment of militant members via social media is a sad part of progress.

2 METHODS

This study used a case study method by observing the experiences of the Muslim millennial generation on social media and other internet resources related to the context of Islamic discourse practices on social media. In addition, the data consisted of descriptions obtained through online and offline interviews with purposively determined informants, comprising of Islamic information services through internet platforms, websites and social media.

We therefore used focus group discussions (FGD) and individual interviews in this research. In addition to our work with young Muslims, we interviewed respondents who played a significant role. Our intention in these interviews was to explore their views about their engagement in social media. In this study, the Muslim millennial generation was represented by the students who had an academic background through Indonesian Higher Education. A number of participants were involved in Islamic religious education from the Faculty of Letters, the Faculty of Sports Science and the Faculty of Social Sciences at Universitas Negeri Malang (145 students).

As stated in Table 1, the respondents were male (60 respondents) and female (85 respondents). From three kinds of faculties, female students dominated the research sample. The mapping of respondents was expected to broaden the scope of research and perform it more comprehensively.

Table 1. Number of the respondents.

Faculty	Male	Female
Faculty of Letters	10	35
Faculty of Social Sciences	10	33
Faculty of Sport Sciences	40	17
Total	60	85

This research is focused on how religious discourses on social media are colored and circulated on establishing the perception on Muslim millennials. This study was conducted throughout 2019–2020. The social media platforms chosen as the focus of this study were Twitter, Instagram and Facebook. Three kinds of social media platforms were chosen because of their highly constant activity in distributing Islamic discourses from the Muslim millennial generation. The data will be elaborated on, in detail, to illustrate the practices of Islamic learning on the Internet in relation to moderatism and radicalism.

3 RESULTS AND DISCUSSION

Based on research in the field using the interview method, it was found that 78% of respondents admitted that they had accessed content that led to a radical understanding of Islam found on social media. Meanwhile, 22% of respondents claimed to have never accessed radical content. From the 78% of respondents who accessed radical content, almost 46% of these students accessed radical content "accidentally." The accident is because respondents often search for keywords to help them in their daily problems and after surfing social media, find themselves "stuck" on websites that contain radical content. Their ignorance led to acceptance of this knowledge and according to the respondents, they found it difficult to filter out this radical knowledge. After accessing the radical content, respondents revealed that the final solution when experiencing "confusion" in responding to the content was to confirm and check the truth of the information with the Islamic Religious Education lecturer in their class. As stated by Monica, one of the students in the English Department, Faculty of Letters, stated:

> "I often use Google to get information about Islamic studies, things that are still confusing, ranging from the law of usury, bombing, jihad and the relationship between Islam and science. The content often leads to portraits of Islam which are "rigid", "cruel" and "sadistic" ... after receiving many explanations like that, I usually meet the Islamic Religious Education Lecturer (PAI) to get confirmation of this."

Therefore, according to the researcher's analysis, one of the functions of Islamic Religious Education (IRE) Lecturers in Public Universities is as a validator of Islamic understandings as experienced directly by students, especially on social media. Lecturers of IRE play a strategic position in efforts to deradicalize the understanding of Indonesian Higher Education's students, moreover, the penetration of social media as a new platform for da'wah further strengthens the challenge of radicalism for the millennial generation. This finding strengthens the idea that educators had a significant influence in controlling moderatism and radicalism in a learning context (Alama 2020).

Furthermore, the mapping of social media's use of respondents is dominated by three platforms: Twitter, Facebook and Instagram.

Table 2 indicates that the respondents have ways of interacting with social media, such as Twitter, Facebook and Instagram. Furthermore, the faculties of Letters, Social Sciences and Sport Sciences have the highest numbers for embracing social media. Facebook had the most users (129 users), followed by Twitter (99 users) and Instagram (104 users).

Related to the previous research on national surveys for religious attitudes in schools and universities in Indonesia conducted by the Center for Islamic and Community Studies (PPIM) UIN

Table 2. The mapping of social media from the respondents.

Faculty	Social Media		
	Twitter	Facebook	Instagram
Faculty of Letters	38	45	39
Faculty of Social Sciences	29	43	35
Faculty of Sport Sciences	32	41	30

Jakarta in 2018 stated that students that do not have internet access have a more moderate opinion compared to those with access to the internet. This fact is quite alarming as 51.1% of student Muslims show a tendency towards intolerance towards minorities. Furthermore, 58.5 percent of student respondents have religious views on radical opinions (Faiqah & Pransiska 2018).

The various responses to student's perceptions on facing the content of moderatism and radicalism are examined below:

Table 3. The mapping of student responses facing moderatism and radicalism.

Faculty	Social Media		
	Confirming with the Lecturer	Browsing Social Media	Asking other students
Faculty of Letters	19	20	6
Faculty of Social Sciences	10	18	15
Faculty of Sport Sciences	12	27	18

Table 3 indicated that respondents have various responses facing moderatism and radicalism in social media. There are three kinds responses: first, confirming with the lecturer (41), second, browsing on social media (75) and thirdly, discussing with other students (39). The use of virtual networks, such as social media as instruments of radicalization and moderation recruitment, will arouse sympathy from individuals who feel the need for solidarity in brotherhood, poverty, social imbalances and political frustration. Browsing social media as an alternative way to learn about Islam needs to be a focus for educators because educators have the responsibility to confirm any questions from students relating to religious issues which are ideally characterized by religious moderation.

This study strengthens previous research (Hilmy 2013; Salem 2016; Arif 2020), that lecturers have the responsibility to assist students in gaining access to information related to Islamic studies with mainstream religious moderation. From the data above, there are quite a lot of students who intentionally or unintentionally access learning content that is indicated to be radical and fundamental, so the role of lecturers must be increased so that students are able to filter this out. The challenge going forward is the need to increase insight and competence, as well as sensitivity for lecturers, so that information relating to religious moderation can become the initial foundation in research and policy development in the relevant government sectors.

In addition, the use of social media as a vehicle for the spread of religious moderation and the ability to use it are important things that must be mastered by lecturers (Herdiansah & Husin 2018; Slama 2018). Lecturers are expected to be able to operate social media and implement them into the framework of mainstreaming religious moderation. This condition is in accordance with the characteristics of the millennial generation who are close to social media, technology and information.

4 CONCLUSION

Discussing the social media and cyberspace and how it influences the Muslim millennial generation is a complicated discourse which involves multiple disciplines. The relationship between social media—as well as technology in general—and religion is essentially reciprocal and interconnected. Online reciprocal relationships are basically intended to describe the connections between the context of Islamic learning practices on the internet and the spectrum of moderatism and radicalism. Under this assumption, social media has become a part of Muslims' daily lives. Islamic knowledge on the internet, as presented in this article, has been demonstrated to have a reciprocal relationship. Optimizing the use of social media as a medium to spread moderatism and to prevent radicalism by targeting the millennial Muslim generation. The development of internet applications, especially on social media, influences the patterns of recruitment and disseminations of radicalism. They are like a multipurpose tool that can be used to teach individuals and groups.

REFERENCES

Alama, M. 2020. A Collaborative Action in the Implementation of Moderate Islamic Education to Counter Radicalism, *International Journal of Innovation, Creativity, and Change*, 11(7), pp. 497–516.

Arif, M. K. (2020). Moderasi Islam (Wasathiyah Islam) Perspektif Al-Qur'an, As-Sunnah Serta Pandangan Para Ulama Dan Fuqaha. *Al-Risalah*, 11(1), pp. 22–43.

Arifianto, A. R. 2019. Islamic campus preaching organizations in Indonesia: Promoters of moderation or radicalism? *Asian Security*, 15(3), pp. 323–342.

Azra, A. 2005. *Islam in Southeast Asia: Tolerance and Radicalism*. Centre for the Study of Contemporary Islam, Faculty of Law.

Baulch, E. and Pramiyanti, A. 2018. Hijabers on Instagram: Using visual social media to construct the ideal Muslim woman. *Social Media+ Society*, 4(4), p. 2056305118800308.

Faiqah, N. and Pransiska, T. 2018. Radikalisme Islam Vs Moderasi Islam: Upaya Membangun Wajah Islam Indonesia Yang Damai. *Al-Fikra: Jurnal Ilmiah Keislaman*, 17(1), pp. 33–60.

Fealy, G. 2012. Islamisation and politics in Southeast Asia: The contrasting cases of Malaysia and Indonesia in *Islam in world politics*. Routledge, pp. 159–176.

Herdiansah, A. G. and Husin, L. H. 2018. Religious Identity Politics on Social Media in Indonesia: A Discursive Analysis on Islamic Civil Societies. *Jurnal Studi Pemerintahan*, 9(2), pp. 187–222.

Hilmy, M. 2013. Whither Indonesia's Islamic Moderatism? A reexamination on the moderate vision of Muhammadiyah and NU. *Journal of Indonesian Islam*, 7(1), pp. 24–48.

Hussain, S. A. and Silcock, B. W. 2019. Social Media Campaign to Improve Religious Tolerance. *Narratives of Storytelling across Cultures: The Complexities of Intercultural Communication*, p. 217.

Irham, I., Haq, S. Z. and Basith, Y. 2020. Deradicalising religious education: Teacher, curriculum and multiculturalism. *Epistemé: Jurnal Pengembangan Ilmu Keislaman*, 15(1), pp. 39–54.

Nur, I. 2020. Embracing Radicalism and Extremism in Indonesia with the Beauty of Islam. *Asian Research Journal of Arts & Social Sciences*, pp. 1–18.

Salem, Z. 2016. The Religion of Social Media: When Islam Meets the Web.

Slama, M. 2017. Social media and Islamic practice: Indonesian ways of being digitally pious. *Digital Indonesia: connectivity and divergence. Singapore: ISEAS Publishing*, pp. 146–162.

Slama, M. 2018. *Practising Islam through social media in Indonesia*. Taylor & Francis.

Suhaimi, S. and Raudhonah, R. 2020. Moderate Islam in Indonesia: Activities of Islamic Da'wah Ahmad Syafii Maarif. *Ilmu Dakwah: Academic Journal for Homiletic Studies*, 14(1), pp. 107–128.

Winarni, L. 2014. The Political Identity of Ulama in the 2014 Indonesian Presidential Election. *Al-Jami'ah: Journal of Islamic Studies*, 52(2), pp. 257–269.

Winarni, L., Agussalim, D. and Bagir, Z. A. 2019. Memoir of Hate Spin in 2017 Jakarta's Gubernatorial Election; A Political Challenge of Identity against Democracy in Indonesia. *Religió: Jurnal Studi Agama-agama*, 9(2), pp. 1–23.

Winarni, L., Yudiningrum, F. R. and Wijaya, S. H. B. 2017. Social Media and the Issue of 'Gafatar' in Indonesia. *KnE Social Sciences*, pp. 115–119.

Woodward, M. 2010. Muslim education, celebrating Islam and having fun as counter-radicalization strategies in Indonesia. *Perspectives on Terrorism*, 4(4), pp. 28–50.

Community Empowerment through Research, Innovation and Open Access – Sayono et al (Eds)
© 2021 Copyright the Author(s), ISBN 978-1-032-03819-3

Sharia accounting perception according to MSMEs in East Java to foster a spirit of business continuity

D. Syariati, D.M. Putri*, S.F. Putri & I.H. Al Siddiq
Universitas Negeri Malang, Malang, Indonesia

Mahirah
University of Malaysia, Kuala Terengganu, Malaysia

ABSTRACT: The aim of this study is to determine whether sharia accounting can be understood and already practiced by MSMEs in East Java Province. This research is a qualitative study that examines the perceptions of sharia accounting in MSMEs, especially in East Java Province. The data was processed from the results of the online focus group discussion (FGD) studies with MSMEs owners. The result shows a positive perception of the owners of MSMEs on the implementation of Islamic accounting. Although MSME owners do not understand the principles of sharia accounting perfectly, they have started to carry out business as stated in sharia accounting. As Muslims, the owners of MSMEs understand sharia accounting can help them carry out buying and selling transactions according to the Al-Qur'an guidelines. Sharia accounting encourages MSME owners to maintain the quality and sustainability of the business.

Keywords: sharia accounting, MSME, business continuity.

1 INTRODUCTION

Indonesia has a high potential for the development of the Sharia industry. With the largest Muslim population, Indonesia has the potential to become the largest consumer and producer of Sharia, especially the halal business has received positive responses from various parties (Aziz & Chok 2013; Battour et al. 2018; Haque et al. 2015; Latif et al. 2014; Mathew 2014; Rezai et al. 2012; Sadeeqa et al. 2013). Thomson Reuters (2018) released the value of spending on halal products in 2017 reached the US $ 2.1 trillion and is estimated to increase to the US $ 3 trillion in 2023. Interestingly, 10% of expenditures in that year were equivalent to Indonesia's spending on halal products (Indonesia Halal Lifestyle Center 2018). Halal food, halal fashion, media and recreation, Muslim friendly travel, halal pharmaceuticals, halal cosmetics, and Islamic finance are several opportunities that give benefits, especially for Micro Small and Medium Enterprises (MSMEs).

Strengthening MSMEs in the halal industry is one of the main strategies in the Indonesian Sharia economic master plan (Ministry of National Development Planning, 43). This is not surprising, considering that 60% of GDP comes from MSME (Micro Small Medium Enterprise) activities. The involvement of MSMEs in this industry has an impact on opportunities for increasing the growth of MSMEs and the halal climate in Indonesia. Responding to this, several MSMEs have positive perceptions of the halal business, including perceptions of non-bank Islamic financial institutions (Baumgartner & Ebner 2010; Barokah & Hanum 2013), crowdfunding sharia (Apriliani et al. 2019), and halal certification (Ahmad & Anwar 2020). A positive perception indirectly emerges from the

---

*Corresponding author: dhika.maha.fe@um.ac.id

DOI 10.1201/9781003189206-22

emerging of Sharia business in Indonesia (Sofyan 2011). On the other hand, Islamic banking institutions' growth has received a negative perception from MSMEs because of the perceived similarity in nature between Islamic banking and conventional banking (Nugroho & Tamala 2018).

If it is studied further, the management of sharia business is different from conventional business. The existence of divine elements is a milestone in business management, including financial management. From an accounting point of view, the use of Sharia accounting is relevant for entities involved in Islamic transactions. Sharia accounting offers different paradigms, principles, and characteristics than conventional accounting (DSAS IAI 2019). Sharia accounting use laws from Al-Quran, As-Sunah, *Ijmak* (ulamas agreement), and *Qiyas* (the similarity of certain events) as a guideline (Khaddafi et al. 2016). Siregar (2015) says that accounting principle according to Islam can be defined as a set of a basic legal and permanent standard, which taken from Islamic Sharia and used by accountants as the reference in doing their work, whether in the entry, analysis, measurement, exposure or explanation and become a reference in explaining certain events. Sharia accounting needs competent human resources in that field (Larasati & Sumardi 2018). Sharia accounting is an alternative accounting. It is normatively conceptualized based on the Islamic principle that life is to return to God, which in Qur'anic term is expressed in an assertion of *innaalillaahiwainnaailaihiraaji-uun* (Triyuwono 2012).

In that case, brotherhood, justice, benefit, balance, and universalism are the principles carried out in the Sharia Financial Accounting Standards. The existence of this principle shows that business activities should aim for the benefit of a broader range. Not only worldly but also spiritual, not only material but also spiritual, not only individually but also collectively. In other words, Islamic accounting is oriented towards business continuity that emphasizes quality, not just profit. With regard to this, the number of Muslim entrepreneurs in Indonesia can become fertile ground for the implementation of sharia accounting to increase the going concern of MSMEs.

Unfortunately, the study about the perception of MSMEs towards sharia accounting has not been massively carried out. So far, studies on the perception of MSMEs have only been carried out on conventional accounting even though there is an enormous potential for the use of sharia accounting as a provision to enter the rapidly growing halal industry. On the other hand, MSMEs need to adjust to an increase in sharia transactions. In addition, the development of sharia accounting is also crucial for MSMEs because it can encourage a potential going concern on the continuity of MSME businesses. This study aims to examine the perception of the MSMEs which are owned or managed by Muslim entrepreneurs towards sharia accounting, especially in East Java Province. East Java Province was chosen because this province has the third larger MSMEs in Indonesia. This study is essential to complete the study of the practice of MSMEs sharia accounting. Moreover, this study's result can become one of the evaluation sources related to the implementation policy of Sharia accounting for MSMEs in Indonesia.

2   METHODS

This research is descriptive qualitative research. Some informants were chosen from each city in East Java and adjusted to the types of their businesses. There were 15 informants selected from the service, trade, and manufacture sectors. Types and sources of data used in this study include primary data and secondary data. Primary data was processed from the results of online focus group discussion (FGD) studies with Micro Small Medium Enterprise (MSMEs) owners. The FGD was conducted online because this research was carried out during the Covid-19 Pandemic. Secondary data in this study consisted of (a) government regulations regarding the MSMEs; (b) government regulations about sharia accounting; and (c) Al-Quran, *As-Sunah, Ijmak*. Secondary research data were obtained from the study of literature through search engines. Data analysis in this study used an interactive model. The components of the interactive model are (1) data collection, (2) data reduction, (3) data presentation, and (4) drawing conclusions. The data validation was carried out through triangulation of sources between data from MSMEs owners.

3 RESULTS AND DISCUSSION

This research was conducted because MSMEs make a significant contribution to the Indonesian economy (Rukiah et al. 2019). Based on the Law of the Republic of Indonesia Number 20 of 2008 concerning Micro, Small, and Medium Enterprises (MSMEs), micro-enterprises are individual productive businesses and/or individual businesses that meet the law's criteria. As described in the law, a small business is a productive business carried out by individuals or business entities. Then, medium-sized enterprises are productive economic enterprises carried out by individuals or business entities with total net assets or annual sales proceeds that meet the law's criteria. Meanwhile, large enterprises are economic enterprises carried out by business entities with a net worth, or annual sales proceeds greater than medium enterprises.

The presence of MSMEs becomes essential for the wheels of the country's economy because MSMEs are a form of business that can provide jobs. Besides, MSMEs' activities can expand employment opportunities and provide broad economic services to the community. MSMEs play an important role in encouraging economic growth, increasing public income equality, and realizing national stability. Indonesian MSMEs contribute significantly to gross domestic product (GDP). Based on the data from the Ministry of Cooperatives, Small and Medium Enterprises (2010-2018), Indonesian MSMEs contributed up to IDR 8,573.9 trillion to Indonesia's GDP (based on current prices) in 2018. Indonesia's GDP in 2018 was IDR 14,838.3 trillion, so the contribution of MSMEs reached 57.8% of GDP. In addition, MSMEs employ 116,978,631 people or 97% of the total Indonesian workforce (MSMEs and Large Units). There are 64,194,057 MSMEs in Indonesia or 99.99% of the total business units in Indonesia.

A large number of MSMEs in Indonesia is accompanied by the fact that Indonesia is a country with the largest number of Muslims in the world, which further strengthens the development of the sharia industry. In this case, the presence of sharia accounting is an answer to the demands related to recording transactions in accordance with the *Al-Quran, As-Sunah, Ijmak* (ulamas agreement), and *Qiya*s (the similarity of certain events) as a guideline. Based on the research results, it is known that there are positive perceptions of the owners of MSMEs towards the implementation of sharia accounting. Even though MSMEs owners do not understand the principles of sharia accounting comprehensively, they have started to carry out their business procedure according to sharia accounting. That procedure includes (a) staying away from usury; (b) set aside zakat for each business's revenues, and; (c) transact with a clear contract with the buyer. As Muslims, the owners of MSMEs are well aware that sharia accounting can help them carry out buying and selling transactions according to the Al-Qur'an guidelines.

In its development, MSMEs have strong resilience in facing an economic crisis. But in its management, MSMEs are constrained on the aspects of management and finance, such as the less quality of human resources who cannot record or bookkeeping the financial condition. Therefore, MSMEs owners really hope that there will be sharia accounting training for daily buying and selling transactions. In addition, MSME owners also expect technical guidelines for buying and selling transactions using sharia accounting principles. The guidelines are containing rules and examples of MSMEs transactions presented in the Islamic accounting standard. These guidelines also explain the terms in Islamic accounting and the time of their use. As a complement, sharia accounting guidelines specifically for MSMEs also contain verses of the Qur'an related to transactions that occur. This guideline is expected to be a window of knowledge and information for MSMEs related to Islamic accounting's comprehensive implementation. In addition, this guideline will also be socialized to MSMEs using regular training and assistance both online and offline. With this assistance, MSMEs will be guided to understand and practice the contents of the guidelines. Through this preparation and assistance, it is expected that MSMEs can improve the quality of financial management through sharia accounting that is oriented towards divine values. Sharia financial reporting applications not only focus on applications that are free of usury or gambling (speculation). More than that, it is expected to provide benefits for the community, especially for MSMEs. With the main values of Islam, shariah accounting aims for justice, holistic welfare, and benefit for all people.

Therefore, the financial report records information about a company in the accounting period that summarizes the company's financial performance. Financial reports are useful for stakeholders in analyzing and interpreting financial performance and the company's condition (IAI 2016). In general, the purpose of both conventional and Sharia financial reports is to present financial reports. However, sharia accounting juxtaposes accounting science with the values contained in the *Al-Quran, As-Sunah, Ijmak (*ulamas agreement*), and Qiyas (*the similarity of certain events). During the reign of Caliph Umar bin Khattab, accounting known as an *alamel, mubashar, al-kateb,* the person responsible for recording and reporting financial and non-financial information. Especially for an accountant by the name *Muhasabah* or *Muhtasib* (Nurhayati 2009). Syahatah (2001) formulated the six purpose of *muhasabah* (accounting) in Islam, namely maintaining the property (*hifz al-amwal*), the existence of *al-kitabah* when there is a dispute, can help in drawing conclusions, determine the results of the business to be deliberate, determine and calculate the rights of the union in business and determine, reward, reply or sanction. Syafrida Hani et al. (2018) said the purpose of sharia financial reports is to provide information concerning the financial position, performance, and changes in the financial position of an Islamic entity that is useful for a large number of users in making economic decisions and other objectives. Those purposes are described below.

a. Improve compliance with sharia principles in every transaction and business activity.
b. Provide information on the compliance of Islamic entities with sharia principles as well as information on assets, liabilities, income, and expenses that are not in accordance with sharia principles.
c. Information to help evaluate the fulfillment of the entity's and sharia's responsibilities towards the trust in securing funds, investing them at a reasonable profit level.
d. Information regarding investment returns obtained by investors and owners of temporary funds and information regarding the fulfillment of obligations. The social function of Islamic entities, including the management and distribution of *zakat, infaq, alms,* and *waqf.*

Sharia Accounting is the science of accounting or accountability of all assets and economic activities of an individual business or group or company following the Qur'an and As-Sunnah to achieve true wealth or prosperity or 'Falah' (Choudhury 2005). The objective of implementing sharia accounting is to achieve socio-economic justice and carry out our worship in fulfilling our obligations to Allah SWT. It is also a form of our responsibility for individual duties in reporting all matters relating to financial reports. Sharia accounting techniques is resulted from accurate accounting information to calculate zakat and accountability horizontally to Allah SWT based on morals, faith, and piety.

In accordance with the description above, the Islamic nuances caused by the implementation of sharia accounting can affect the way MSME owners conduct business. MSME owners feel that sharia accounting can help them keep their business away from things that are not following religious teachings. Thus, MSMEs owners feel calmer in life because they work not only for money but also for the pleasure of Allah. In addition, with the existence of sharia accounting practices, every incoming and outgoing transaction is strictly maintained. Sharia accounting indirectly encourages MSME players to maintain their business quality, from the financial perspective, production to services. Sharia accounting can inspire someone, to be honest, and trustworthy. By adhering to quality, honesty, and trust, a business entity's sustainability can be maintained properly.

The feeling of calm arising from the integration of divine values with knowledge through sharia accounting can be called a part of celestial management. Putri (2020) said that celestial management is a management science based on the interpretation of spiritual values and principles derived from the word of Allah to be implemented in life. It is the belief that work is part of worship. Work is not only related horizontally to humans but also vertically upwards to the Allah SWT. Through celestial management, MSMEs owners can conduct continuous improvement for their organizations. However, business is not just a matter of material but also a missionary field that can reassure us.

4 CONCLUSION

In accordance with the description above, the Islamic nuances caused by the implementation of sharia accounting can affect the way MSMEs owners conduct business. MSMEs owners feel that sharia accounting can help them keep their business away from things that are not following religious values. Thus, MSMEs owners feel calmer in life because they work not only for money but also for Allah's pleasure. With the existence of sharia accounting practices, every incoming and outgoing transaction is strictly maintained. Sharia accounting indirectly encourages MSME players to maintain the quality of their business. By adhering to quality honesty and trust, a business entity's sustainability can be maintained properly. The importance of MSMEs' existence is especially evident during the critical economic times after the Covid-19 pandemic. MSMEs is a sector that can maintain the stability of people's income. Business management that juxtaposes knowledge with faith in Allah, is an effort to survive critical times without violating ethics.

REFERENCES

Ahmad, B., & Anwar, M. K. 2020. Analisis Respon Perilaku Usaha Minuman Kopi (Coffe Shop) Terhadap Kewajiban Sertifikasi Halal. *Jurnal Ekonomika Dan Bisnis Islam, 3*(2).

Apriliani, R., Ayunda, A., & Fathurochman, S. F. 2019. Kesadaran Dan Persepsi Usaha Mikro Dan Kecil Terhadap Crowdfunding Syariah. *Amwaluna: Jurnal Ekonomi Dan Keuangan Syariah, 3*(2), 267–389.

Aziz, Y. A., & Chok, N. V. 2013. The Role Of Halal Awareness, Halal Certification, And Marketing Components In Determining Halal Purchase Intention Among Non-Muslims In Malaysia: A Structural Equation Modeling Approach. *Journal Of International Food & Agribusiness Marketing, 25*(1), 1–23.

Barokah, S., & Hanum, A. N. 2013. Analisis Presepsi Nasabah Dan Perkembangan Umkm Setelah Memperolah Pembiayaan Mudharabah (Studi Kasus Bprs Binama Kota Semarang). *Maksimum, 3*(2).

Battour, M., Hakimian, F., Ismail, M., & Boğan, E. 2018. The Perception Of Non-Muslim Tourists Towards Halal Tourism. *Journal Of Islamic Marketing.* 4(1), 15–25.

Baumgartner, R. J., & Ebner, D. 2010. Corporate Sustainability Strategies: Sustainability Profiles And Maturity Levels. *Sustainable Development, 18*(2), 76–89. Https://Doi.Org/10.1002/Sd.447

Choudhury, M. A. 2005. Islamic Economics And Finance: Where Do They Stand? *The Islamic Quarterly, 49*(4), 247–280.

Haque, A., Sarwar, A., Yasmin, F., Tarofder, A. K., & Hossain, M. A. 2015. Non-Muslim Consumers' Perception Toward Purchasing Halal Food Products In Malaysia. *Journal Of Islamic Marketing.* 4(1), 15–25.

Khaddafi, M., Siregar, S., Harmain, H., Nurlaila, Zaki, M., & Dahrani. 2016. *Akuntansi Syariah Meletakkan Nilai-Nilai Syariah Islam Dalam Ilmu Akuntansi* (A. Ikhsan (Ed.); 1st Ed.). Penerbit Madenatera.

Larasati, M., & Sumardi. 2018. Pelatihan Akuntansi Lembaga Keungan Syariah Bagi Siswa Di Smk Plus Ashabulyamin Kabupaten Cianjur. *Jurnal Inovasi Sosial & Pengabdian Kepada Masyarakat, 1*(2), 168–177. Https://Doi.Org/10.22236/Syukur

Latif, I. A., Mohamed, Z., Sharifuddin, J., Abdullah, A. M., & Ismail, M. M. 2014. A Comparative Analysis Of Global Halal Certification Requirements. *Journal Of Food Products Marketing, 20*(Sup1), 85–101.

Mathew, V. N. 2014. Acceptance On Halal Food Among Non-Muslim Consumers. *Procedia-Social And Behavioral Sciences, 121*, 262–271.

Nugroho, L., & Tamala, D. 2018. Persepsi Pengusaha Umkm Terhadap Peran Bank Syariah. *Jurnal Sikap (Sistem Informasi, Keuangan, Auditing Dan Perpajakan), 3*(1), 49–62.

Nurhayati, S. 2009. *Akuntansi Syariah Di Indonesia*. Penerbit Salemba.

Putri, D. M. 2020. Efforts To Realize Halal Business: Implementation Of Celestial Management In Msmes. *Kne Social Sciences, 2020*, 185–195. Https://Doi.Org/10.18502/Kss.V4i9.7325

Rezai, G., Mohamed, Z., & Shamsudin, M. N. 2012. Non-Muslim Consumers' Understanding Of Halal Principles In Malaysia. *Journal Of Islamic Marketing.*

Rukiah, Nuruddin, A., & Siregar, S. 2019. Islamic Human Development Index Di Indonesia (Suatu Pendekatan Maqhasid Syariah). *Istinbáth, 18*(2), 307–327.

Sadeeqa, S., Sarriff, A., Masood, I., Farooqi, M., & Atif, M. 2013. Evaluation Of Knowledge, Attitude, And Perception Regarding Halal Pharmaceuticals, Among General Medical Practitioners In Malaysia. *Archives Of Pharmacy Practice*, 4(4).

Sofyan, R. 2011. *Bisnis Syariah, Mengapa Tidak?: Pengalaman Penerapan Pada Bisnis Hotel*. Gramedia Pustaka Utama.

Syahatah, H. 2001. Pokok-Pokok Pikiran Akuntansi Islam. Terj. *Khusnul Fatarib. Jakarta: Akbar Media Eka Sarana*.

Triyuwono, I. 2012. *Akuntansi Syariah: Perspektif, Metodologi Dan Teori Pt Rajagrafindo Persada*. Jakarta.

Community Empowerment through Research, Innovation and Open Access – Sayono et al (Eds)
© 2021 Copyright the Author(s), ISBN 978-1-032-03819-3

Infographic development of Blambangan Kingdom for history learning in senior high school

M.N.L. Khakim*, I.Y. Afhimma, K.A. Wijaya, M.R.I. Ardiansyah, & Marsudi
Universitas Negeri Malang, Malang, Indonesia

ABSTRACT: Teaching materials become an essential component of learning. History learning is often considered synonymous with boring material. The purpose of this study is to develop interesting teaching materials for the historical learning. The development of teaching materials used historical research data. The Blambangan Kingdom (1293–1772) was one of the important ancient kingdoms in East Java that needed to be taught in schools. The method of this study was research and development. The steps of research and development were literature study, product design, product specification, design validation, product manufacturing, initial product testing, product revision, extensive product testing, and dissemination. Infographic teaching materials become a solution for history learning. This study develops an infographic teaching materials with a focus on the discussion about the history of Blambangan Kingdom in senior high school. This infographic is up to date because it used an attractive cartoon figure, a concise explanation, and an additional QR code to enrich the information. The implementation of infographics on 35 respondents indicates that infographic teaching materials on the history of Blambangan Kingdom are effectively used in learning supported by material points, instructions for use and evaluation questions.

Keywords: teaching materials, Infographic, Blambangan Kingdom, history learning

1 INTRODUCTION

Learning in the current era has progressed, especially history learning. History learning can be carried out in two approaches, namely chronologically and thematically (Sapto et al. 2019). In fact, history learning is often considered boring and dry because it only talks about facts and concepts. Of course, teachers need to gain facility related to the preparation of alternative learning in media and teaching materials to support the teaching and learning process. In optimizing the learning process, an atmosphere that can activate interaction between educators and students, students and students, and students with learning materials are needed (Rokhmah 2017). The existence of teaching materials can foster students' interest so that it creates interactive interactions and teachers have used various kinds of teaching materials in history learning. The example of those materials is textbooks, modules, handouts, worksheets, models or mock-ups, audio teaching material,s and other interactive teaching materials about Indonesia's ancient history. Researchers have also conducted research on the ancient history of Indonesia, such as ancient bathing sites in Pasuruan (Khakim et al. 2020). A study about Hindu-Buddhist buildings has also been carried out by researchers at Selokelir Temple (Khakim et al. 2020). This research is in line with previous studies on the history of ancient Indonesia for teaching history.

One of the teaching materials in the modern era is infographic teaching materials. According to Barnes in Resnatika et al. (2018), infographics are visual data that aims to provide information from

---

*Corresponding author: moch.nurfahrul.fis@um.ac.id

DOI 10.1201/9781003189206-23

a phenomenon where readers can interpret the infographic's meaning. The infographic teaching materials were developed because, so far, teaching materials have become an obstacle in history learning. Besides, it can visually package information so that readers can remember and understand without having to read long texts. Infographic becomes an effective alternative teaching material for history learning because the explanation of the material is in the form of points so that messages or information can be conveyed. Often, the teacher's teaching materials are not attractive, so that learning becomes boring, and learning objectives cannot be achieved. Thus, this study develops infographic teaching materials to make history learning interesting and not boring. The history of Blambangan Kingdom 1293–1772 AD in senior high school was selected as the material because this history is not attractive to be discussed. Thus, the history of the Blambangan Kingdom seems to have been eliminated from the national historiography. Therefore, it is necessary to have a historical discussion about the history of the Blambangan Kingdom 1293–1772 AD in infographic teaching materials to attract student interest in learning.

2 METHODS

This research used the research and development method with a model of Borg & Gall. The procedure of research and development methods were literature study, product design, product specification, design validation, product manufacturing, initial product testing, product revision, extensive product testing, and dissemination (Setyosari 2016). The first step was a literature study to collect the information of Blambangan kingdom and identify the problem about the lack of learning material about the history of Blambangan kingdom in senior high school. The second step was the product design, carried out through designing types of teaching materials suitable for teaching the history of the Blambangan Kingdom. The third step was product manufacturing to compile the infographic about the history of the Blambangan Kingdom. The fourth step was initial product testing by the teaching materials expert. Meanwhile, in the fifth stage, the product was revised based on the review's results by the teaching material expert. The sixth step was extensive product testing to 35 students in senior high school. This step also involved a google form filled by the first grade of senior high school. In the seventh stage, the teaching material was distributed to history teachers via the internet. A quantitative approach was to analyze data based on the google form result. Thus, the last step was making a conclusion about how effective this teaching material to teach about the history of the Blambangan Kingdom for student in senior high school.

3 RESULTS AND DISCUSSION

3.1 *A brief history of Blambangan Kingdom*

The Blambangan Kingdom was located at the eastern tip of the island of Java, as the last Hindu kingdom in Java. Based on the source of the Balambangan inscription published by Prabu Jayanegara, Balambangan has been named a self-reliant area because it has been instrumental in suppressing the rebellion Nambi and Blambangan and were designated as a farming area (Hardiati et al. 2008). Based on the information of the Balambangan inscription, the Blambangan Kingdom had existed since the era of the Majapahit Kingdom. However, this inscription is not a marker of the establishment of the Blambangan Kingdom. Many historians even believe that the Blambangan Kingdom coincided with the fall of Singhasari and the establishment of the Majapahit Kingdom in 1293 AD (Margana 2012). When the Majapahit Empire collapsed, the Blambangan Kingdom still survived.

According to Babad Blambangan, the first King of the Blambangan Kingdom was Menak Sopal, but his reign was not explained. According to the Babad Sembar, Lembu Miruda was the first king of Blambangan and made Watu Putih Panarukan the capital of the Blambangan Kingdom. This kingdom controled most of the eastern tip of the island of Java, namely Banyuwangi, Jember,

Lumajang, Bondowoso, and Situbondo. According to Lekkerkerker, the Blambangan was the Hindu Kingdom with a strong Hindu fortress and could not be conquered by the Mataram Sultanate.

After the collapse of Majapahit, the Demak Sultanate conquered Tuban in 1527 AD, Wirasari in 1528 AD, Surabaya in 1531 AD. Besides, Lamongan, Blitar, and Wirasaba had their turn in 1541 AD and 1542 AD. In 1546 AD, the Demak Sultanate tried to conquer Blambangan but failed despite its success in expanding hegemony to Hindu areas in the interior of Java (Anshori & Dri Arbaningsih 2008).

For almost three centuries, the Blambangan Kingdom was under the influence of Mataram and Bali's supremacy struggle. The kingdom of Mengwi, Bali was then more successful in exerting influence in Blambangan. Bali's kings used the Blambangan Kingdom to fight the expansion of the Mataram Sultanate (Margana 2012). The Balinese Kingdoms did this because they were afraid that Islam could penetrate Bali and change Bali's cultural structure. Therefore, Blambangan Kingdom was used as the last stronghold of Hinduism in Java so that Islam did not spread to the island of Bali. Blambangan had important trading centers (Veenman et al. 1927). It was clear that the VOC's main interest in Blambangan was to impose a trade monopoly by the VOC on the territory of the Blambangan Kingdom.

Blambangan fighters staged a counterattack against the VOC. The attacks were aimed at the VOC strongholds, carried out simultaneously and completely. The VOC troops were defeated by the Blambangan fighters. Lieutenant Kornet Tinne and countless VOC soldiers were killed. The remaining VOC soldiers withdrew to Kuta Lateng, then retreated to Ulu Pangpang (Ali 2002). The Blambangan fighters won a brilliant victory. Early in 1772 AD, the VOC mobilized the East Java coastal regents to attack Blambangan. They decided to burn down the village that had the food supplies for the Blambangan warriors in Bayu. On April 20, 1772 AD, about 200 Madurese soldiers led by Alap-Alap and 112 Javanese soldiers from Besuki led by Captain Wayan Buyung attacked the Blambangan fighters. The VOC sent a further military expedition to conquer Bayu. On May 16, 1772 AD VOC troops and their allies attacked Bayu and were again defeated by Blambangan fighters (Utojo 2015).

The VOC sent a further military expedition to conquer Bayu. On June 11, 1772 AD, there was another battle between the VOC and Blambangan. In this battle, the leader of the resistance, Rempeg Jogopati died (Margana 2012). Although the resistance leader died, the founders of Bayu decided to wage a Puputan war, an all-out war against the VOC. VOC prepared the final blow to break the resistance of the Blambangan fighters in Bayu. The VOC was sure that with the killing of Rempeg Jogopati the Blambangan fighters would not stem the attack. After this incident, the condition of the Blambangan Kingdom, which was completely destroyed, stopped the struggle to continue. After October 11, 1772 AD, the Puputan War launched by the Blambangan Kingdom ended.

3.2 *Implementation of the infographic for history learning*

The study of history has a wide range of material, one of which is the Hindu-Buddhist kingdom's material. The material was derived from competency standards, namely analyzing the journey of the Indonesian nation during the traditional countries, then becoming basic competencies (KD) 3.6. The material on the Hindu-Buddhist kingdom in Indonesia is taught in senior high school in relation to this. Based on this material, there are teaching materials that support the learning activities. Infographic teaching materials are one of the visual teaching materials that provide a concept in simplifying information or learning materials in the form of images, maps, and graphics to make it easier for readers. These teaching materials are arranged systematically according to the basic competencies of history learning, especially on Hindu-Buddhist main subjects. Hindu-Buddhist infographic teaching materials were implemented at senior high school students in grade X semester 1. Hindu-Budha's main material is focused on the presentation and analysis of material on the history of the Blambangan Kingdom (1293–1772).

The preparation of this infographic teaching materials also considered the effectiveness and inno-vation in the application of learning so that the material presented reaches the students. Infographic

Table 1.　Infographic teaching materials on the history of Blambangan Kingdom 1293–1772 AD.

Design infographic	The preparation of infographic teaching materials with the presentation of material on the history of Blambangan Kingdom 1293–1772 AD based on essential points supported by instructions for use and evaluation questions accompanied by a barcode containing material that has been compiled briefly and densely with references. • Page 1: Brief explanation of the material • Page 2: Instructions for using infographic teaching materials 1. The teacher provides directions for students to prepare cell phones. 2. The teacher provides a brief explanation as an overview of the material. 3. The teacher shows the infographic teaching materials to the students. 4. Teachers form groups/individuals to analyze the points presented in the infographic teaching material. 5. The teacher provides a barcode containing material on the Hindu-Buddhist history of the Blambangan Kingdom 1293–1772 AD. 6. The teacher provides the opportunity for students to ask questions and answer questions. • Page 3: Evaluation questions

teaching materials have the advantage of explaining material from complex to simple and practical to use. In that sense, it is in the form of points that cover the entirety so that it is easy to understand and adjusts the learning needs of students to suit learning objectives. The existence of this infographic teaching materials and the evaluation of the question are expected to support students' learning interest and understanding in addition to books and explanations from the teacher. The implementation of the material on the history of Blambangan Kingdom 1293–1772 AD by combining infographic teaching materials can properly become qualified teaching materials and support student learning outcomes through visualization of infographics (Table 1).

After the presentation of the statement of the infographic teaching materials on the history of the Blambangan Kingdom presented in Table 1, data research on the effectiveness of the teaching material was collected. In this case, 35 respondents filled out a questionnaire on the google form. The research results are presented in percentage form in the Table 2.

Based on the questionnaire data, most of the responses to questions numbered one to four are "Effective" in assessing infographic teaching materials with the history of the Blambangan Kingdom 1293–1772 AD material. Respondents were presented with four questions related to the effectiveness of infographic teaching materials as supporting learning history. The results obtained from the research data show that an average of 40% of each question item refers to an effective result. The respondents also suggested various criticisms and suggestions, such as image addition, changing fonts, choosing colors, and arranging to be more innovative. Therefore, infographic teaching materials can be classified as effective teaching materials used in learning based on data from 35 respondents, which are equipped with instructions for use and evaluation questions to support students' cognitive aspects.

Table 2. Data on the results of research on the effectiveness of infographic teaching materials.

No.	Question indicator	Descriptions of the Effectiveness Scale				
		Ineffective	Less effective	Effective enough	Effective	Excellent
1.	A	0,0%	2,9%	37,1%	40%	20%
2.	B	2,9%	11,4%	28,6%	40%	17,1 %
3.	C	0,0 %	0,0%	25,7 %	42,9 %	31,4%
4.	D	0,0 %	0,0%	34,3 %	42,9%	22,9%

Description from question indicator

A How effective the infographic teaching materials in presenting the history of the Blambangan Kingdom?

B How effective the infographic teaching materials in the history of Blambangan Kingdom 1293–1772 AD material used for the independent learning process?

C How effective the infographic teaching material in the history of Blambangan Kingdom 1293–1772 AD material used for the learning process accompanied by teachers?

D How effective the Infographic teaching material in the history of Blambangan Kingdom 1293–1772 AD in increasing your knowledge of Hindu-Buddhist material?

3.3 Infographic teaching material development

The existence of teaching materials eases the educational aspect. According to the National Center for Competency Based Training (Prastowo 2015), teaching materials are all materials used to assist teachers in carrying out the learning process. Teaching materials are all materials that are systematically arranged to show all the competencies used in the learning process to achieve learning objectives. Teaching materials contain the entire learning technique designed by the teacher and implemented in the learning process. Teaching materials are made based on curriculum needs and student conditions and must be according to the times (Krismawati et al. 2018).

In the reality of education, there are still many educators who use conventional teaching materials; however, it is not enough that the current teaching materials must be compiled in an attractive, effective, and efficient manner by developing their innovations. Teaching materials must be arranged in an attractive manner so that students have more interest in historical learning material. One of the distributions of teaching materials according to their form (visual) is printed teaching materials. Printed teaching materials are teaching materials in the form of paper that function as learning needs and convey information and examples of printed teaching materials, one of which is an infographic (Prastowo 2015).

Infographic stands for Information and Graphics. Infographics are part of the science that supports business and commerce, which are presented in the form of visual information and contain more text and image data to be more attractive. The design of teaching materials with graphic elements has a higher value (Aldila 2019). Infographics use visual elements such as images, charts, maps, and charts so that the message on the infographic can be conveyed systematically and easily understood by readers. The visual form of the infographic is designed as attractive as possible so that the teaching materials used by teachers and students are not boring. Infographics have the benefit of conveying information clearly and effectively (Yahya et al. 2017).

Students are one of the audiences who experience limited time in utilizing various written historical sources. Infographics are a solution for students who want to dig as deep as possible knowledge with limited time and go hand in hand with a visual learning style. It can be in the form of diagrams or maps that will make it easier for students because the human brain will more easily understand the learning material presented (Sari 2017). The delivery of material equipped with visual illustrations can provide a stimulus for the brain to imagine history. With the existence of imagination, history learning becomes fun and as if students experience historical events themselves. This is because historical imagery will make history learning livelier and more enjoyable. The imaginative side of history that is raised will make students construct a historical event (Subakti 2010).

The advantages of infographic teaching materials are that it exposes complex data to simple. The messages presented in an image are faster to understand, arouses imagination, is an objective source of information, and explains events coherently. Correct understanding of the learning messages conveyed is a necessity for the teacher to achieve learning objectives. In making infographics, certain design principles must be considered, such as simplicity, coherence, emphasis, and balance. Visual elements that need to be considered are shapes, lines, space, texture, and color (Sari 2017). The visualization generated from the infographic is an integral part of conveying information so that it is easy and fast to understand. The novelty of this infographic is it used the attractive cartoon figure and additional QR code. The steps for using this infographic-based teaching material are, first, the teacher introduces the importance of studying the history of the Blambangan kingdom in East Java. Second, the teacher displays this infographic on the History of the Blambangan Kingdom. Third, students read and understand the history teaching materials of the Blambangan Kingdom. Fifth, students scan the QR code on the infographic to explore information about the Blambangan Kingdom. Fourth, students work on evaluation questions under the guidance of the teacher. Finally, students and teachers discuss the answers to the evaluation questions and conclude the history lesson of the Blambangan Kingdom.

4 CONCLUSION

Teaching materials are all materials arranged systematically, according to student needs, basic competencies, and learning objectives. Infographic teaching materials are teaching materials based on visual graphic design. The Blambangan Kingdom (1293–1772) was founded almost simultaneously with the Majapahit Kingdom's establishment. The Blambangan Kingdom lost to the VOC expeditionary force, so that it had to be destroyed. The material on the history of the Blambangan Kingdom is suitable to be taught with infographic teaching materials. This study examines the uses of infographic teaching materials on the historical material of the Blambangan Kingdom in senior high school. Based on data from 35 respondents, the infographic teaching materials are effective to be used in learning supported by material points, instructions for use, and evaluation questions. This infographic's novelty is it used an attractive cartoon figure, the information is easy to understand, and an additional QR code.

REFERENCES

Aldila, T. H., Akhmad Arif Musadad, & Susanto. (2019). Infografis Sebagai Media Alternatif Dalam Pembelajaran Sejarah Bagi Siswa SMA. *Andharupa: Jurnal Desain Komunikasi Visual & Multimedia Vol. 05 No. 01*, 4.

Ali, H. (2002). *Sekilas Perang Puputan Bayu Sebagai Tonggak Sejarah Hari Jadi Banyuwangi Tanggal 18 Desember 1771.* Banyuwangi: Pemerintah Kabupaten Banyuwangi.

Anshori, N., & Dri Arbaningsih. (2008). *Negara Maritim Nusantara Jejak Sejarah Yang Terhapus.* Sleman: Tiara Wacana.

Babad Tawang Alun. (n.d.). 2–5.

BudiSantoso. (2006). Sriwijaya Kerajaan Maritim Terbesar Pertama di Nusantara. *Jurnal Ketahanan Nasional, XI (1)*, 5.

Hamid, A. R. (2013). *Sejarah Maritim Indonesia.* Yogyakarta: Penerbit Ombak.

Hardiati, E. S., Djafar, H., Soesoro, & Ferdinandus, P. &. (2008). *Sejarah Nasional Indonesia II: Zaman Kuno.* Jakarta: Balai Pustaka.

Krismawati, N. U., Warto, & Suryani, N. (2018). Analisis Kebutuhan pada Bahan Ajar Penelitian dan Penulisan Sejarah di Sekolah Menengah Atas (SMA). *Briliant: Jurnal Riset dan Konseptual Volume 3 Nomer 3.*

Kuntowijoyo. (2013). *Pengantar Ilmu Sejarah.* Sleman: Tiara Wacana.

Margana, S. (2007). *The Struggle for Hegemoni of Blambangan.* Leiden: Universiteit Leiden.

Margana, S. (2012). *Perebutan Hegemoni Blambangan.* Yogyakarta: Pustaka Ifada.

Marihandono, D., & Harto Juwono. (2008). *Sultan Hamengkubuwono II Pembela Tradisi dan Kekuasaan Jawa.* Yogyakarta: Banjar Aji Production.

Muljana, S. (2008). *Sriwijaya*. Bantul: LKiS Pelangi Aksara Yogyakarta.

Nurmaria. (2017). Gerakan Sosial Politik Masyarakat Blambangan Terhadap Kompeni di Blambangan Tahun 1767–1768. *Patanjala Vol. 9 No. 2*, 3.

Poesponegoro, M. D., Nugroho Notosusanto, R.P. Soejono, & R.Z. Leirissa. (2010). *Sejarah Nasional Indonesia Volume 2*. Jakarta: Balai Pustaka.

Prastowo, A. (2015). *Panduan Kreatif Membuat Bahan Ajar Inovatif*. Jogjakarta: Diva Press. Cetakan viii.

Resnatika, A., Sukaesih, & Kurniasih, N. (2018). Peran Infografis sebagai Media Promosi dalam Pemanfaatan Perpustakaan. *Jurnal Kajian Informasi & Perpustakaan Vol.6 No.2*.

Rokhmah, F. N. (2017). Pengembangan Bahan Ajar Sejarah Kebudayaan Islam "Masa Rasulullah Periode Mekah" berbasis Accelerated Learning. *Tesis. Progam Studi Pendidikan Agama Islam Pascasarjana Institut Agama Islam Negeri Purwokerto*.

S., H. B. (2006). Sriwijaya Kerajaan Maritim Terbesar Pertama di Nusantara. *Jurnal Ketahanan Nasioal, (XI)1*, 5.

Sapto, A., Lutfiyah, A., Ridhoi, R., & Khakim, M. N. (2019). Pengembangan Kajian Sejarah Tematik Sebagai Alternatif Bahan Ajar Sejarah Tingkat Menengah Atas di Blitar. *Jurnal Praksis dan Dedikasi Sosial Vol. 2 No.1*.

Sari, E. P. (2017). *Pengembangan Media Berbentuk Infografis Sebagai Penunjang Pembelajaran Fisika SMA Kelas X*. Lampung: FTK Universitas Islam Negeri Raden Intan (Skripsi).

Sholeh, K. (2017). Jalur Pelayaran dan Perdagangan Sriwijaya Pada Abad ke-7 Masehi. *Siddhayatra Vol. 22*, 2.

Sholeh, K. (2019). Pelayaran Perdagangan Sriwijaya dan Hubungannya Dengan Negeri-Negeri. *Jurnal Historia Volume 7, Nomor 1*, 2.

Subakti. (2010). Paradigma Pembelajaran Sejarah Berbasis Konstruktivisme. *Paradigma Pembelajaran Sejarah. ....(Y.R. Subakti) SPPS, Vol. 24, No. 1*, 3.

Sudjana, I. M. (2001). *Nagari Tawon Madu: Sejarah Politik Blambangan abad XVIII*. Bali: Larasan Sejarah.

Sundoro, H. M. (2008). *Pangeran Rempeg Jagapati: Pahlawan Perjuangan Kemerdekaan di Tanah Blambangan*. Banyuwangi: Dinas Kebudayaan dan Pariwisata Kabupaten Banyuwangi.

Suparman, H. (1987). *Bahasa Osing di Kabupaten Banyuwangi*. Jakarta: Universitas Indonesia.

Utojo, S. (2015). *Kumpulan Catatan Sejarah Kabupaten Banyuwangi*. Banyuwangi: Yayasan Puri Gumuk Merang.

Veenman, H., Zonen, & Wageningen. (1927). *Balambangansch Adatrecht*. Banyuwangi: Pusat Study Budaya Banyuwangi.

Wahyudi, B. S. (2013). *Pengembangan Bahan Ajar Berbasis Model Problem based Learning Pada Pokok Bahasan Pencemaran Lingkungan Untuk Meningkatkan Hasil Belajar Siswa Kelas X Sma Negeri Grujugan Bondowoso*. Jember: FKIP Unej.

Yahya, S., Wibawa, M., & Surya, A. (2017). Infografis Kompetensi Multimedia dan Desain Grafis di Provinsi Jawa Timur. *Journal of Art, Desaign, Art Education and Culture Studies (JADECS) Vol. 2 No. 2 (journal2.um.ac.id)*.

Yuliati. (2014). Kejayaan Indonesia Sebagai Negara Maritim (Jalesveva Jayamahe). *Jurnal Pendidikan Pancasila dan Kewarganegaraan, Th. 27, Nomor 2*, 6.

Community Empowerment through Research, Innovation and Open Access – Sayono et al (Eds)
© 2021 Copyright the Author(s), ISBN 978-1-032-03819-3

Carrying capacity of local communities to developing Tamansari Tourism Village, Banyuwangi

A. Purnomo*, G.S.A. Wardhani, I.H.S. Buddin, M. Rahmawati, P. Glenn, Idris, &
B. Kurniawan
Universitas Negeri Malang, Malang, Indonesia

ABSTRACT: This research aims to determine local communities' carrying capacity towards developing Tamansari tourism village in Licin District, Banyuwangi. This tourism village is a tourist attraction located on the border of the Banyuwangi district. Its geographic conditions are located at the foot slope of Mount Ijen with a distinct culture of the plantation community. The development of a tourism village was driven by a village business entity that assists the community in developing tourism objects. This study used a quasi-qualitative research method by direct interviews aimed at the representative of the Village Owned Enterprises (BUMDes), small business owners, and Tamansari tourism village managers. Interviews for data collection were used to ensure the information from local communities and managers. The results show that several strategies can be applied in developing the potential of the Tamansari village, such as 1) developing cultural tourism, culinary and natural charm, 2) involving the local community in developing Tamansari tourism village. Therefore, some recommendations for this purpose are developing cultural, natural, and culinary tourism, improving public facilities, developing accessibility, and strengthening collaboration among stakeholders.

Keywords: carrying capacity, village tourism, tourism development

1 INTRODUCTION

The development of tourism cannot be separated from the participation of local communities. The local community plays a role in developing the village's potential (Butler, 1980). The community around tourist objects can make various economic and tourism activities through tourism, for example, accommodation, services (transportation, information), food stalls, tourist objects, and others (Lee & Back, 2006). These activities can increase people's income and reduce the unemployment rate.

Tourism development is managed by the village government in the concept of a tourist village. A tourism village program will provide practical benefits to improve the people's standard of living in it. As stated in the Regulation of the Minister of Tourism of the Republic of Indonesia Number 29 of 2015, the development of rural-based tourism (tourism villages) will drive economic tourism activities in rural areas, preventing the urbanization of rural communities to cities. (Choi & Murray, 2010). The development of rural tourism will encourage the preservation of nature, which will reduce global warming. The tourism village program is expected to be able to make a positive contribution to community development.

This study examines the support and role of the community in developing tourism villages. A tourist village is not a village created only for tourists, but a village designed to present a unique and exciting culture so that it is also attractive to tourists (Putra & Pitana, 2010). Besides, a tourist

---

*Corresponding author: agus.purnomo.fis@um.ac.id

village functions to reintroduce ancient cultures and traditions that are not yet known by the current generation (Bowers, 2016; Lee & Jan, 2019; Olsder K&M, 2006).

2 METHODS

Tamansari Banyuwangi Village is located on the southern slope of the Ijen mountain complex. The Ijen Volcano area was initially a sizeable volcanic body that experienced eruptions on a large scale to form an oval caldera basin-oriented Southwest – Northeast. After the big eruption, smaller eruptions occurred through Raung, Pendil, Rante, and Merapi's volcanoes. The material from the eruptions of the Raung, Pendil, Rante, and Merapi volcanoes covered most of the body of Mount Ijen and left the northern slopes not covered in new material. The whole process can be observed.

This study used several instruments to collect data, namely, observation, interviews, and documentation. Direct interviews were conducted with one of the representatives from BUMDes, small business owners, and the Tamansari tourism village manager. Interviews for data collection were used to confirm the information from local communities and managers. The interview adopted a semi-structured approach to questioning. This interview form was identified as the most suitable for examining the predetermined results from observational notes and qualitative analyzes (Newman, 2017). The framework method was used to analyze qualitative data in the form of an interview transcript (Brunt, 1997) and the selection of sources according to the work structure (Krippendorf, 1987) based on the respondent's involvement with tourism.

3 RESULTS AND DISCUSSION

Tamansari Banyuwangi Village is located on the southern slope of the Ijen mountain complex. The Ijen Volcano area was originally a large volcanic body that experienced eruptions on a large scale so that it formed an oval caldera basin-oriented Southwest – Northeast. After the big eruption, smaller eruptions occurred through Raung, Pendil, Rante, and Merapi's volcanoes. The material from the eruptions of the Raung, Pendil, Rante, and Merapi volcanoes covered most of the body of Mount Ijen and left the northern slopes not covered in new material.

The Ijen Caldera is an exotic area unit in the Ijen Volcano Area. The Ijen Caldera is a potential regional unit for various ecotourism purposes, such as sulfur utilization culture, Ijen Crater natural tourism, tea plantation tourism, hiking, and forest and savanna Tourism at the top of a volcano. Ijen Caldera management needs to pay attention to the ongoing volcanic processes not to cause environmental damage.

The currently visible Ijen Volcano Complex was generated through a long series of processes. The process of forming the Ijen Volcano Complex can be traced by observing the current condition of surface morphology. The difference in surface morphology that is formed today indicates a difference in the process of formation from the beginning to the present.

The surface morphology of the Ijen Caldera that is currently visible is in the form of the escarpment on the north side and a series of volcanic cones on the south side. The material for the escarpment on the north side has an older material age compared to the slopes on the south side, which are composed of volcanic cone slopes. The incision pattern that occurs on the north slope surface looks coarser than the incision pattern that appears on the surface of the slope in the South, which shows differences in the period of formation.

A tourism village program is a form of community participation in developing potential in the village by utilizing the original rural environment while increasing agricultural potential through a supporting tourism program. Most of the residents of Tamansari Bayuwangi Village residents generally still rely on the informal sector, such as agriculture. In contrast, there are plantation managers and entrepreneurs whose management is under the auspices of BUMDes or known as Village-Owned Enterprises in the formal sector. Tamansari Village is a village with an agricultural area and a nature-based tourism area with one of the Osing Banyuwangi culture's superiorities.

Tamansari tourist village presents natural beauty as a tourist attraction that can be enjoyed. Besides, there are several other potentials such as homestays, tourist transportation facilities, guide services, and several small and medium enterprises managed by BUMDes Ijen Lestari. The Head of Desa Tamansari states that this institution had managed 53 inns belonging to Tamansari Village people. It also helps innkeepers with promotions via online travel agents. The village with a typical Osing nuance has a tourist attraction that becomes interest to tourists, namely the Osing culture. This Banyuwangi regular Osing cultural performance is performed when there are significant events in the village by presenting cultural performances presented in the village's open space.

BUMDes Ijen Lestari has also expanded its wings to tour packages in Tamansari Tourism Village, referred to as Dewi Tari. Not only visiting Mount Ijen, but tourists who come can also be invited to enjoy tourist objects in the village such as Kampung Bunga and Kampung Susu and Kampung Miner and enjoy coffee produced by the village community. Tourists can also explore the pine forest and learn local wisdom from the community. The idea of managing a tour package emerged because Tamansari Village has only been passed by tourists when visiting Mount Ijen. Dewi Tari tour packages allow tourists to stop by and enjoy the many tourist objects available. Tourist visits to the village will provide economic benefits to residents (Butler, 1980).

The tourism village as an activity that involves the local community's participation is intended to realize community independence to live well by utilizing the potential of a village, including Tamansari village in Licin Banyuwangi District. Community involvement aims to contribute to developing and developing promising areas to protect the environment (Lee & Back, 2006).

A tourism village is a form of community empowerment by integrating attractions, accommodation, and supporting facilities that are presented in a structure of community life that integrates with the prevailing procedures and traditions (Nalayani, 2016). Tamansari Village, which is the main entrance to Ijen Crater via Banyuwangi, is transformed into a tourist village. Each visitor is charged a fee of IDR. 5,000 for a single visit; this fee includes retribution and travel insurance.

Community is also involved in providing tourism services in the form of a shelter or culinary. Villagers are required to register if they want to open a guest house to serve visitors who will climb to Ijen Crater. However, visitors to this tourist village are more focused on enjoying Banyuwangi's typical Osing culture.

The Osing cultural products offered by Dewi Tari are manifested in various types such as traditional arts, music festivals, and Osing specialties. This art is a traditional event that is routinely held in the Tamansari village. The types of arts that still survive today include Gandrung, Barong *Ider Bumi*, *Kebo-Keboan*, and *Seblang*. Gandrung art is the Osing Banyuwangi tribe's original art, which is still popular today—infatuated with being the mascot for tourism in Banyuwangi Regency. Gandrung is a dance that embodies joy, intimacy, and community togetherness through dance (Indiarti, 2013). Of the many traditions and customs in Banyuwangi, the *Gandrung* and *Kebo-keboan* dance is an early and typical Banyuwangi tradition that is still often performed today in several areas of Banyuwangi by the Osing tribe.

The flexibility of community involvement has resulted in the community's awareness of improving the quality of tourism services. The BUMDes coordinator conducts a comparative study to open new insights about product development. Visiting guests, such as students or tourism village managers, also benefits from sharing management information (Son & Pitana, 2010; Lee & Jan, 2019). Outreach and outreach to the community resulted in small and medium enterprises assisted by BUMDes, such as setting up small shops in the Tamanasari rest area.

The BUMDes collaborates by cooperating with the private sector to improve the economy of the community. Using CSR funds from private companies, several tourist sites in the village were developed, starting from the milk village (Dusun Ampel Gading), the village of flowers (Dusun Jambu), culinary businesses, creative industries, and services.

The development of the diversity of tourism products increases the attractiveness of tourists. The essential thing in a destination must be the attraction that can attract tourists' attention both psychologically and in real terms (Hermantoro, 2011). Tourism development, especially in developing tourism destinations, is part of a plan to advance and improve the local area's real conditions to contribute and add value and benefit local communities around tourist areas, tourists, and local governments.

Previous studies show that tourism is an effective way to overcome poverty (Croes, 2014; Lepp, 2007). The economic benefits obtained by the community trigger them to get involved. Tourism also generates various sources of livelihood (Lepp, 2007; Diedrich & García-Buades, 2009), such as opportunities to provide tourism services/products.

The pattern of society that is easy to imitate in providing tourism services/products in the first stage creates competition. However, if managed properly, they will become a healthy competition. The benefits of having a community are the dissemination of benefits from tourism activities that are obtained. The community also plays a role in strengthening the carrying capacity of the social or physical environment (Lee & Jan, 2019). The findings show that tourism activities increase the number of local people concerned with preserving the social and physical environment. They realize that the environment around where they live is an attraction for tourists (Lee, 2013). The impact is that the profits they get are partly invested in improving environmental quality (Uysal, et al., 2012; Hunt & Stronza, 2014). Therefore, at this stage, managers do not feel significant economic benefits. The economic benefits they get are prioritized for quality improvement investments that aim to increase tourist visits (Uysal, et al., 2012).

4 CONCLUSION

Tamansari tourism village is a form of tourism development that involves the participation of local communities to realize community independence to live well by utilizing a village's potential. The local community's carrying capacity is needed to strengthen cooperation between village communities and allow local people to be creative in producing tourism product innovations that have economic value. Besides, the development of tourist objects through the village can also provide added value in introducing an area to the general public quickly.

REFERENCES

Bowers. 2016. Developing sustainable tourism through ecomuseology: A case study in the rupununi region of Guyana. *Journal of Sustainable Tourism, 24*(5), 758–782.

Brunt, & Courtney. 1999. Host perceptions of sociocultural impacts. *Annals of Tourism Research, 26*(3), 493–515.

Brunt, P. 1997. *Market Research in Travel and Tourism.* Oxford: Butterworth-Heinemann.

Butler. 1980. The concept of a tourist area cycle of evolution: Implications for management of resources. *Canadian Geographer, 24*(1), 5–12.

Choi, & Murray. 2010. Resident attitudes toward sustainable community tourism. *Journal of Sustainable Tourism, 18*(4), 575–594.

Croes. 2014. The role of tourism in poverty reduction: An empirical assessment. *Tourism Economics, 20*(2), 207–226.

Diedrich, & García-Buades. 2009. Local perceptions of tourism as indicators of destination decline. *Tourism Management, 30*(4), 512–521.

Hermantoro, H. 2011. *Creatif – Based Tourism (Dari Wisata Rekreatif Menuju Wisata Kreatif).* Yogyakarta: Galang Press.

Hunt, & Stronza. 2014. Stage-based tourism models and resident attitudes towards tourism in an emerging destination in the developing world. *Journal of Sustainable Tourism, 22*(2), 279–298.

Indiarti, W. 2013. Pengembangan Program Desa Wisata dan Ekowisata Berbasis Partisipasi Masyarakat di Desa Kemiren. *Banyuwangi: Badan Perencanaan Pembangunan Daerah Kabupaten Banyuwangi.*

Krippendorf, J. 1987. *The Holiday Makers: Understanding the Impact of Leisure and Tourism.* Oxford: Butterworth-Heinemann.

Lee. 2013. Influence analysis of community resident support for sustainable tourism development. *Tourism Management, 34*, 37–46.

Lee, & Back. 2006. Examining structural relationships among perceived impact, benefit, and support for casino development based on 4 year longitudinal data. *Tourism Management, 27*, 466–480.

Lee, & Jan. 2019. Can community-based tourism contribute to sustainable development? Evidence from residents' perceptions of the sustainability. *Tourism Management, 70*, 368–380.

Lepp. 2007. Residents' attitudes towards tourism in Bigodi village, Uganda. *Tourism Management, 8*, 876–885.

Nalayani. 2016. Evaluasi Dan Strategi Pengembangan Desa Wisata Di Kabupaten Badung, Bali. *Jurnal Master Pariwisata, 2*(2), 189–198.

Newman, L. 2017. *Metodologi Penelitian Sosial: Pendekatan Kualitatif dan Kuantitatif* (7 ed.). Jakarta: Indeks.

Olsder K, & M, V. d. 2006. *Destination Conservation: Protecting Nature by Developing Tourism.* Netherlands: IUCN National Committee of The Netherlands.

Putra, D., & Pitana, I. G. 2010. *Pariwisata Pro Rakyat.* Jakarta: Kementerian Kebudayaan dan Pariwisata.

Sastrohadi, J., Sianturi, R. S., Rahmadana, A. D., Maritimo, F., Wacano, D., Munawaroh, … Pratiwi, E. S. 2014. *Bentang Sumberdaya Lahan Kawasan Gunungapi Ijen dan Sekitarnya.* Yogyakarta: Pustaka Pelajar.

Uysal, Woo, & Singal. 2012. The tourist area life cycle (TALC) and its effect on the quality-of-life (QOL) of destination community. Dalam *Handbook of tourism and qualityof-life research* (hal. 423–443). Netherlands: Springer.

Vargas-Sánchez, Plaza-Mejia, & Porras-Bueno. 2009. Understanding residents' attitudes toward the development of industrial tourism in a former mining community. *Journal of Travel Research, 47*(3), 373–387.

Community Empowerment through Research, Innovation and Open Access – Sayono et al (Eds)
© 2021 Copyright the Author(s), ISBN 978-1-032-03819-3

Development of iconic spot replications in Kampoeng Heritage Kajoetangan as learning media for indische empire culture for tourists

L. Sidyawati*, J. Sayono, S.D. Anggriani & M.N.L. Khakim
Universitas Negeri Malang, Malang, Indonesia

J.K.B. Ali
Universiti Teknologi Mara, Malaysia

ABSTRACT: One of the Indische Empire-style cultures left from the Dutch era whose existence can still be seen and studied and whose artifacts are yet passed down from generation to generation in Malang City is Kampoeng Heritage Kajoetangan. This village is located on Jalan Jend, Basuki Rachmat Gg. VI, Kauman, Kec. Klojen, Malang City, East Java. Situated in an alley in a heritage building area, this village is one of the tourist villages frequented by children, adolescents and adults. It has houses from the Dutch colonial era and is occupied by their owners. This village consists of three hamlets, each of which has iconic assets, namely *Rumah* Namsin, *Rumah* Jacoeb, and *Kali* Sukun. These icons need to be introduced to the younger generation as part of the history of Malang City. To make it easier to recognize the icons, researchers developed a replica using the Borg and Gall development model, which simplified the process into four steps taking into account the time available, namely Research and Information Collecting, Planning, Developing the Preliminary Form Product and Final Product Revision. The results of this research are (1) the replica of the interior of the *Rumah* Namsin, (2) the replica of the exterior of the *Rumah* Namsin, (3) the replica of the interior of the *Rumah* Jacoeb, (4) the replica of the exterior of the *Rumah* Jacoeb and (5) Replica of the *Kali* Sukun and Jembatan Semeru.

Keywords: replica, iconic, *kampoeng* heritage kajoetangan, learning media, indische empire, culture

1 INTRODUCTION

Culture is a way of life formed from many elements such as knowledge systems, language, technology and equipment systems, art systems, livelihood systems, religious systems, and social systems in a group of people or society. Selo Soemardjan and Soeleman Soemardi (1964) state that culture is everything that is created by humans both tastes and creations. Culture is inherited from generation to generation and an inseparable part of human life. In historical records, the occurrence of inherited culture from generation to generation can be investigated by tracing back the various legacies that are still left.

One of the cultures whose existence can still be seen and studied and the artifacts that are yet passed down from generation to generation in Malang City is Kampoeng Heritage Kajoetangan. This village is located in the heart of the city, precisely on Jalan Jendral Basuki Rachmat Gg. VI, Kauman, Klojen, Malang City, East Java. This village is one of the tourist villages frequently visited by children, adolescents, and adults. It has houses from the Dutch colonial era occupied

---

*Corresponding author: lisa.sidyawati.fs@um.ac.id

DOI 10.1201/9781003189206-25

by their owners. The interview results conducted by the researcher, with the head of the Tourism Awareness Group (Pokdarwis), reveal that there were 60 old houses identified in this village. The houses are relatively preserved in their original forms.

After the researchers conducted field observations on 60 houses identified in the Kampoeng Heritage Kajoetangan, the houses were classified to have an Indische Empire style. Indische Empire style is the influence of Dutch colonial architectural styles in Indonesia from the mid-18th century to the early 19th century. This style was lifted from French architecture, which was later brought by Governor-General Herman Willem Daendels to Indonesia. The application of the Indische Empire style in Indonesia was adapted to climat conditions and the availability of materials (Handinoto & Soehargo 1996).

However, the field's problem is that visitors can take selfies in front of the house or inside, but the educational content is still lacking. There are no relevant learning materials regarding the cultural reconstruction of Kampoeng Heritage Kajoetangan until they became the relics we see today. The results of interviews with tourists also discover that sometimes they are not always able to enter the house and see the inside details. Tourists are also no longer able to reconstruct how the house was once arranged and intended. Therefore, researchers feel the need to develop learning media through the replication of this heritage building so that later visitors who come here and take selfies in front of or inside the house can also learn about the reconstruction of the past of how this house was used.

2 METHODS

Developing a replica of a house icon in Kampoeng Heritage Kajoetangan as a learning medium for Indische Empire culture for this tourist followed the Borg & Gall development model. The reason for choosing the Borg and Gall (2003) development model was that the steps in this study follow the initial concept of development that the researcher had carried out. Sukmadinata (2006) explains that the Borg and Gall development model contains ten steps for implementing a research and development strategy. In this study, it was simplified into four phases, considering the limited time of the research. Those phases included research and information collecting. At this stage, the researcher made observations in the Kampoeng Heritage Kajoetangan to collect data to identify problems in the tourist village to find solutions to visitors' questions. Besides this, researchers also collected data about heritage buildings used as replicas for the learning materials. The second stage was planning, where the researcher arranged work steps and alternative solutions if problems in preparing the application occurred, along with the solutions to the issues. The third stage was to develop a preliminary form of the product. At this stage, the researchers measured and created a 3D building design that would be realized as a replica and developed an iconic replica of the Kampoeng Heritage Kajoetangan building. In this stage, the replica was made with a specific scale precisely the same as the original building. During the final product revision, the speakers' last input of results was refined to become the finished product.

3 RESULTS AND DISCUSSION

Kampoeng Heritage Kajoetangan is an area designated as a cultural heritage area by the government of Malang City. This village tour offers cultural tourism that contains historical education by showing the Dutch colonial heritage house's architecture in the Indische Empire style, which is still maintained until today. Indische means like Indies or Indies. Indischgast or Indischman, in Dutch, means a Dutch person who used to live in Indonesia for a long time. Hij is Indisch means he has Indonesian blood. Indisch culture is a mixture of European, Indonesian, and specific Peranakan Chinese culture (Milone 1966), which emerged primarily as an architectural expression in the mid-18th and 19th centuries. It was originated from European bachelors living in Indonesia and their concubinage or marriage to Indonesian, Eurasian or Chinese women. However, the prototype of this culture grew due to the relationship between European men who took the mistresses of Indonesian female domestic servants (Nyai) and formed a family (Milone 1966).

Building architecture, outdated equipment, or other items are also available in this area, such as *ontel* bicycles, cooking utensils, lamps, windows, cameras, telephones and other home furnishings. Kayutangan village also keeps many remains of past civilizations in the form of shopping buildings, the tomb of Eyang Honggo Kusumo, the Tandak cemetery, the Krempyeng market, Dutch irrigation channels, waterways, "thousand stairs" and other points of high historical value in Malang (Khakim 2019). The houses and colonial heritage buildings in the Indische Empire style were transformed into replicas. It serves as one of the learning media for tourists so that tourists not only go around the village and take selfies but can also get information about the reconstruction of the buildings. It also explains its main function when the building was used. With this learning media, tourists can imagine seeing themselves in the past and living with the Kayutangan people in colonial times. Gerlach and Ely (1971) said that the media, when understood in broad terms, is human, material, or events that build conditions that enable students to acquire knowledge, skills or attitudes. Media is used in the field of teaching or education so that the term becomes educational media or learning media. While students, in this case, are tourists visiting Kampoeng Heritage Kajoetangan. Furthermore the benefits of learning media are also stated by Sudjana and Rivai (2002), that using teaching media can enhance student learning in teaching, which in turn is expected to increase the learning outcomes they achieve.

The researcher only took three iconic heritage buildings representing three hamlets to be replicated due to research time limitations. The selected building icon represents the economic system, technology systems, equipment and the ordinary social systems in the Dutch colonial era. Those heritage icons are: (1) *Rumah* Namsin, (2) *Rumah* Jacoeb, and (3) the *Kali* Sukun irrigation channel flanked by the Semeru Tunnel in the north and Talun in the south.

3.1 *Replica of the interior and exterior of the Rumah Namsin*

Rumah Namsin is located at Jalan Basuki Rahmad No. 31, Malang City. The logo of the Kampoeng Heritage Kajoetangan was taken from this house building. *Rumah* Namsin is estimated to have been founded in the 1900s, and the first owner was a Dutchman named V. Doorene. In 1924, *Rumah* Namsin was purchased by a Dutch national named L. C. Verhey, and this building was used as a motorcycle dealership under the Indian, Harley, Douglas and F.N brands and sold parts for Ford cars. When the Japanese came to Indonesia, the Verhey family left Indonesia and returned to

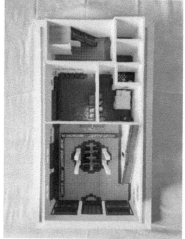

Figure 1.   Replica of the Interior of the 1st Floor of the *Rumah* Namsin.

Figure 2.   Replica of the Interior of the 2nd Floor of the *Rumah* Namsin.

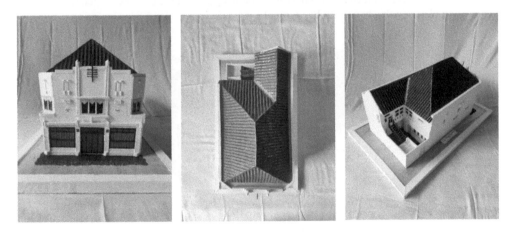

Figure 3.   Replication of *Rumah* Namsin's Exterior.

Rotterdam in the Netherlands. The Japanese took over *Rumah* Namsin as a tribute deposit office. After Indonesia's independence in 1950, this house was bought by the Namsin family, and used it as a Singer sewing machine shop and for the production of ice lollies. In the 1975's, they sold the house through a broker named Wen Zhen. After Wen Zhen, this house was bought by Siho Ismanto (Liem Zhong Hoo). This house was repaired and returned to its original form.

In the 1978, this house was finally restored and began to be occupied again. Since then, this house has been used as a private residence and is not used for trading. The furniture inside is also obtained from Siho's colleagues, ranging from chairs and tables to cupboards of various sizes. In 2006, Siho fell ill. This house was vacated because he and his wife were old and had difficulty climbing stairs to the 2nd floor. On March 18, 2011, Siho Ismanto died in a hospital in Surabaya. He was finally buried in Kepanjen cemetery, Malang. His wife named Lilik Tamiati (Tan Mee Swen), eventually moved to a house located at Jalan K.H. Hasyim Ashari No. 30, Malang.

In 2015, Siho Ismanto's son named Suyono (Liem Ting Soen), planned to restore this building's heritage atmosphere. This house was painted and refurbished. In the middle of 2016, this house began to be used once a week on Fridays for Divine Services. Since then, the *Rumah* Namsin has become known by the residents of Malang City as a historical house, until today this house has always been controlled every day. The replica sizes for this *Rumah* Namsin are 46.8 cm in length, 27 cm in width, and 35 cm high. The tools and materials used to manufacture these replicas included plywood, paperboard, wood glue, cork, mica plastic, cardboard, acrylic paint, brushes, cutters and scissors.

3.2 *Replica of interior and exterior of the Rumah jacoeb*

Rumah Jacoeb is one of the old buildings in Kampoeng Heritage Kajoetangan, which was still maintained in its original form (structure and some furniture) until now. *Rumah* Jacoeb is located at Jl. AR. Hakim II No. 1193. It was built around 1920, with the owner Jacoeb, who was a Padangnese who bought a Javanese house in Kayutangan in 1930. Jacob was a level 1 draftsman in the government who also had a hobby of painting. The results of his paintings are still firmly displayed on the walls in his house. When viewed from above, the shape of this house building is a triangle. This house has an area of 113 m2 with a building area of 60 m2. There has never been any change from the beginning until now, including the "thump" in front of the house. The Jacoeb family used the "thump" that looks like a horse saddle as a piece of furniture to relax and chat in the afternoon with neighbors. The lower "saddle" is a chair, and the upper "saddle" is a table where coffee, tea, and snacks are placed. This house was initially gray, but the owner changed it to brown. The table and chairs in this house are Vanderpool Limensen.

Figure 4.   Replication of *Rumah* Jacoeb's Interior.

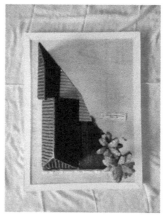

Figure 5.   Replication of *Rumah* Jacoeb's Eksterior.

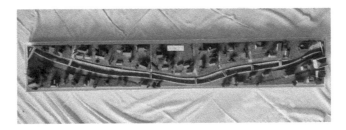

Figure 6.   Replica of the *Kali* Sukun and Semeru Tunnel.

The replica sizes for this *Rumah* Jacoeb are in 40 cm length, 29 cm width, and 20 cm high. The tools and materials used to manufacture these replicas include plywood, paperboard, wood glue, cork, mica plastic, cardboard, acrylic paint, brushes, cutters and scissors.

3.3   *Irrigation channel*

Malang City is an area on the bend of the Brantas River, where during the Dutch era, a fort was built on the left side of the Brantas River and is now called the Klojen area (Ridjal et al. 2016). The irrigation channel around the Kampoeng Heritage Kajoetangan comes from a river break centered in the Oro-oro Dowo area. The irrigation channel was built during the reign of the Dutch East Indies and is still being maintained today. There are about six river shards in the underground passage and are still active today. One of the river fragments in the Kampoeng Heritage Kajoetangan area, named *Kali* Sukun, also flows to the Kebon Agung sugar factory in Malang and flows to the rice fields around the Sawahan area (currently Sawahan Petrol Station). Not many things have been changed from the *Kali* Sukun irrigation channel in the Kampoeng Heritage Kajoetangan. The current situation is being repaired to avoid flooding because of the river's shallowness (H. Udin). This *Kali* Sukun is located around the Talun, Kauman, and Kasin areas (Schaik 1996).

Inside the Kampoeng Heritage Kajoetangan, there is an irrigation channel. The irrigation channel became a barrier between the colonial housing area (Jl. Tengger, Jl. Dorowati, Jl. Arjuna) and indigenous housing. Indigenous housing is arguably older than colonial housing. Indigenous housing was also formerly owned by the colonialists, who then often changed its ownership. Colonial housing is assumed to have existed since Bouwplan V (the 1920s), which focused on development in the Talun area. During those years, European immigrants arrived, requiring the government to provide them with a place to live. It is even more iconic because there are two tunnels at both ends of the channel in the Kayutangan area, namely the Semeru tunnel in the north and the Talun tunnel on the south side. The two tunnels support the road bridge in Malang City, whose building structure is still maintained today.

This irrigation canal's replica sizes are in 110 cm length, 18 cm in width, and 13 cm high. The tools and materials used to manufacture these replicas include plywood, paperboard, wood glue, cork, mica plastic, cardboard, acrylic paint, brushes, cutters and scissors.

4   CONCLUSION

This replica of the iconic spot in the Kampoeng Heritage Kajoetangan is a learning medium for learning about the Indische Empire culture for tourists that provides educational information. This replica has succeeded in modernizing the delivery of information about cultural artifacts of colonial heritage, which initially could only be seen from the outside and became selfie spots. It has become a replica where tourists can reconstruct past events from when these spots were still used for activities during the colonial period. In the future, the replica of the iconic spot in Kampoeng Heritage Kajoetangan will be used by other researchers with more detailed information and equipped with more innovative technologies.

REFERENCES

Borg & Gall. 2003. *Education Research*. New York: Allyn and Bacon.

Frampton, Kenneth.1985. *Modern Architecture A Critical History*. Thames And Hudson Ltd: London

Gerlach, Vernon S. & Donald P. Ely. 1971. *Teaching & Media: A Systematic Approach*. Second edition. New Jersey: Prentice Hall, Inc.

Handinoto. 2009. Indische Empire Style: Gaya Arsitektur "Tempo Doeloe" yang Sekarang sudah Mulai Punah. *Jurnal Dimensi*, 3(1), 35–45.

Handinoto & Soehargo, P. H. 1996. *Perkembangan Kota & Arsitektur Kolonial Belanda di Malang*. Yogyakarta: ANDI.

Khakim, Nurfahrul Lukmanul. 2019. Urgensi Pengelolaan Pariwisata Kampung Heritage Kajoetangan Malang. *Jurnal Teori dan Praksis Pembelajaran IPS*, 4(2) 15–22.

Milone, Pauline D. 1966a. *Qeen City of the East: Metamorphoses of a Colonial Capital, Berkley*. University Of California. Phd: Thesis.

Milone, Pauline D. 1966b. Indische Culture And Its Relationship to Urban Life. *Jurnal Comparative Studies In Society & History*, Vol. 9, 407–426.

Perdana, Andini. 2019. Naskah La Galigo: Identitas Budaya Sulawesi Selatan di Museum La Galigo. Pangadereng: *Jurnal Hasil Penelitian Ilmu Sosial Dan Humaniora*, 5 (1), 25–38.

Ridjal, Abraham Mohammad. 2016. Building Form berdasarkan Sejarah Kawasan Bangunan pada Jalan Basuki Rahmat Malang. *Jurnal RUAS*, Vol. 14 No 2. 1–15.

Seminar Nasional Seni Dan Desain: "Reinvensi Budaya Visual Nusantara". Jurusan Seni Rupa dan Jurusan Desain Universitas Negeri Surabaya.

Soemardjan, Selo & Soelaeman, S. 1964. *Setangkai Bunga Sosiologi*. Jakarta: Lembaga FE-UI.

Sudjana, Nana & Rivai, Ahmad. 2002. *Media Pengajaran*. Bandung: Sinar Baru OFFSET

Sukmadinata, Nana Syaodih, 2006. *Metode Penelitian Pendidikan*. Bandung: PT. Remaja Rosdakarya.

Van Schaik, A. 1996. *Malang*. Beeld van een Stad. Purmerend Asia Maior.

Community Empowerment through Research, Innovation and Open Access – Sayono et al (Eds)
© 2021 Copyright the Author(s), ISBN 978-1-032-03819-3

Community-based tourism: Capability and community participation in tourism development

Idris*, A. Purnomo & M. Rahmawati
Universitas Negeri Malang, Malang, Indonesia

ABSTRACT: Community-based tourism (CBT) is a term that has various interpretations, which in its implementation has led to a long debate among scholars and practitioners. Community involvement in tourism development, which is a crucial issue in CBT, is often questionable and often cannot be implemented in reality. Some of the reasons for this include a lack of understanding and practical experience with CBT. This paper aims to discuss the implementation of CBT, particularly as an instructive review of the CBT literature to synthesize key elements regarding the community's capabilities and participation in the development of tourism destinations. Various literatures show that local communities' knowledge and ability in developing tourist destinations is an effective key in supporting the implementation of CBT for sustainable tourism development. In addition, the community's capability to participate in tourism development directly requires the attention of stakeholders and government officials so that it can be a key support for community-based tourism development.

Keywords: CBT, capability, community participation, tourism development

1 INTRODUCTION

As the idea of sustainable development began to develop in tourism, increasing attention was paid to the issue of involving local communities in destination development policies and planning (Dangi & Jamal 2016; Dodds et al. 2018; Ernawati et al. 2017). However, many researchers and practitioners still question the value and application of public participation theory despite the developing field. They claim that the theory is too naive and expensive to implement (Kim & Kang 2020; Lee & Jan 2019; Lindstrom & Larson 2016). Part of the problem is that the definition of sustainability in tourism tends to be too general and all-embracing for its practical implementation (Mayaka et al. 2018). Furthermore, the lack of practical actions that promote and test community engagement strategies is also seen as a weakness (Curcija et al. 2019).

Community participation is an integral part of sustainable tourism development, especially in accordance with community-based traditions. In addition, it also highlights the right to be involved in the transformation of a community into a tourist destination and the benefits that may be derived from this involvement (Lindstrom & Larson 2016). Several attempts have been made to articulate the practical action of community involvement in tourism development. This is often claimed to be an unrealistic and utopian strategy because of its complexity and high transaction costs (Zielinski et al. 2018).

Discussions about CBT have become prominent, including a mixed understanding of what CBT entails and a critical evaluation of the extent to which it is community-oriented (Dangi & Jamal 2016; Mtapuri & Giampiccoli 2016; Ruiz-Ballesteros & Cáceres-Feria 2016). Despite the acceptance of community-based tourism and the involvement of local stakeholders in tourism development in

---

*Corresponding author: idris.fis@um.ac.id

theoretical debates, effective implementation is still considered a challenge among scholars, as well as practitioners, in the field (Kala & Bagri 2018; Manyara & Jones 2007). Therefore, this article contributes to the debate and implementation of community-based tourism, highlighting the community's participation and capabilities in theoretical tourism destination development.

2   COMMUNITY PARTICIPATION IN TOURISM

Participation is defined as a thought or action taken by a person to contribute to achieve common goals and be responsible for the efforts made (Sastropoetro 1986). Sulistiyorini (2015) describes that participation can be defined as participation or participation by individuals or groups, both material and non-material. While community participation in tourism means the involvement of the community in the process of identifying a problem and tourism potential in their environment, selecting solutions in solving problems, implementing in solving problems, and being able to evaluate them are necessary (Ramadhan 2014).

There are three elements of the community concept of participation in tourism, including forms of responsibility, a willingness to contribute to achieving common goals, and a willingness to be involved in groups (Sulistiyorini 2015). Participation should be based on a will and awareness from within itself so that there will be no coercion element. In addition, participation must be carried out with an awareness of the responsibility for what it does to achieve goals in a group (Sastropoetro 1986).

Community participation in tourism, in general, consists of two types, namely the real form participation and abstract form participation (Mayaka et al. 2018). Sastropoetro (1986) describes the concept of community participation in tourism based on its type classified into three dimensions. First, thought participation is the involvement that is given in the form of ideas, ideas, or constructive thinking. Second, energy participation, which is physical participation in order to achieve the success of a plan. Third, material participation is the involvement of a person or group in the form of money, property, or goods to achieve joint efforts.

The participation of local communities in various tourism activities in the surrounding environment brings various benefits, especially in the economic sector (Kala & Bagri 2018; Lee & Jan 2019; Purnomo et al. 2020). This benefit is what motivates the community to get involved in tourism. The availability of space that can be managed by the community in the tourism sector results in a diversity of livelihoods (Mtapuri & Giampiccoli 2016; Ruiz-Ballesteros & Cáceres-Feria 2016; Zielinski et al. 2018). Management that is applied by local communities in providing various products or services in tourism activities can continue to develop even without government support. In this context, tourism development is managed based on the interaction between local communities and tourists in providing tourism products or services that can provide experiences in exchanging knowledge (Purnomo et al. 2020; Wahyuningtyas et al. 2020).

Local people who have an easy pattern of imitating the provision of services or tourism products at an early stage (exploration) can lead to competition. However, if they can be managed properly, they will become a strong group or community. The existence of a community can provide benefits from tourism activities obtained by the community. In addition, the community also plays a role in strengthening the carrying capacity of the physical or social environment (Lee & Jan 2019). Tourism activities can increase the number of local people who are increasingly concerned with the preservation of the physical and social environment around them because they realize that tourism in the surrounding environment has an attraction for tourists. The impact is that local people get benefits that can be used to improve environmental quality. Therefore, at this stage, tourism managers cannot experience significant economic benefits because the economic benefits are prioritized for quality improvement, which aims to increase the number of tourist visits to existing tourist objects (Purnomo et al. 2020; Wahyuningtyas et al. 2020).

In the future, an increase in the number of tourist visits has implications for the efforts made by the manager in developing various attractions. This stage is called the involvement community stage, in which local community involvement emerges because they already understand the benefits

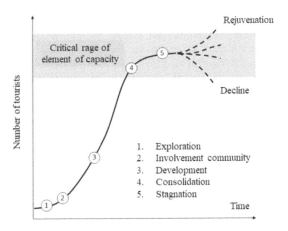

Figure 1. Tourism development stages adapted from Butler (1980).

they get. At this stage, many people build knowledge, both positive and negative, regarding the tourism sector's existence in their surroundings (Butler 1980; Ngo et al. 2018). Positive knowledge is related to the economic benefits that people can feel. Meanwhile, negative public knowledge can be formed from the perceived concern for security, personal life that can be disturbed, or health concern where this can occur when tourists carry infectious diseases while on tourist visits in the surrounding environment. The government has a role in minimizing the occurrence of internal conflicts due to differences in knowledge of local communities (Mayaka et al. 2018). In addition, the government also plays a role as control through regulation or increasing the ability of tourism managers to provide tourism products or services. The regulations that are formed aiming to control any negative impacts from tourism activities must be based on a study of the carrying capacity of the physical and social environment (Lee & Jan 2019).

3 COMMUNITY CAPABILITIES IN CBT

CBT is a local community-based tourism development that acts as a policymaker and receives benefits from the existence of a tourist destination where tourism development will be the basis for sustainable tourism development (Dangi & Jamal 2016; Okazaki 2008; Razzaq et al. 2012). The essence of sustainable tourism ensures that natural, social, and cultural resources in tourism development can be felt by future generations. Thus, CBT is directed as a form of local community involvement in making decisions for the tourism development process that is able to prosper local communities through the benefits obtained (Mtapuri & Giampiccoli 2016).

Through CBT, tourism development can stimulate business activities to generate benefits in various fields such as the economic, social, and cultural fields. When tourism can develop properly, the presence of tourism can provide various benefits that can be felt by people involved in tourism activities, including increasing community income through increasing job opportunities. Some of the work opportunities that the community can take advantage of are opening stalls, becoming ticket officers, souvenirs, and other benefits (Dodds et al. 2018; Lee et al. 2013; Lee & Jan 2019; Pemayun & Maheswari 2017; Utama 2017). In addition, tourism development will also create infrastructure development that can provide potential access for local people to various facilities (Wulaningrum 2018).

The success of developing a tourist attraction can be determined by support from the government or institutions, good and effective management, and active participation from the community and tourism management groups (Purnomo et al. 2020; Wahyuningtyas et al. 2020). Tourism development through CBT focuses on the involvement of local communities (Amilia et al. 2020).

This shows that local communities' capability in managing existing tourism is a major factor in whether tourism objects are able to develop or not. Capability is the ability, expertise, competence, and capacity possessed by individuals or groups to manage potential together, solve problems, and design goals to be achieved together (Dodds et al. 2018; Kala & Bagri 2018; Noho 2014). Community capabilities are assets and attributes that are integrated into society and can be used by them to improve their lives (Bonewati et al. 2017; Manyara & Jones 2007).

Sofield (2003) explains that CBT is a form of tourism development that has the characteristics of tourism that provides opportunities for local people to be involved in tourism development both in terms of planning and implementing it. In addition, people who are actively involved and passively involved will benefit from this tourism business. In addition, CBT is concerned with planning tourism development based on the needs of local communities so that people can enjoy the benefits (Okazaki 2008). There are three principles in CBT, including local communities' involvement in making decisions, local people getting benefits, and education or assistance for local communities related to tourism. This aims to prepare them to have the ability to manage potential existing destinations (Lee & Jan 2019).

Local communities have a role as providers of the main components of tourism in its management. This role cannot be separated from their capabilities related to tourism in the surrounding environment. The main features in tourism include attractions, amenities, and accommodation (Inskeep 1991; Kim & Kang 2020; Pemayun & Maheswari 2017; Zielinski et al. 2018). Attractions are essential elements that attract tourists (Lee & Jan 2019). Tourist attractions are related to the concept of what tourists can see and do in a tourist destination (Setyanto 2019). Attractions are significant in attracting tourists to revisit a tourist destination. Tourist attractions at certain destinations usually cannot be imitated by other destinations because each destination has its own characteristics related to its attractions (Curcija et al. 2019; Dangi & Jamal 2016; Kala & Bagri 2018). Therefore, an attraction should have uniqueness and characteristics that can distinguish it from other destinations.

The capability of local communities in tourism management can also be seen from the availability of supporting facilities. Supporting facilities are needed to facilitate tourists to enjoy existing tourist attractions. Facilities that must be provided include amenities and accommodation (Lawson 1998). Communities that are capable of various tourism activities provision are one of the crucial components that must be in tourism management (Razzaq et al. 2012). Increasing community capability is one of the driving forces in the development of existing tourism. This shows that in community-based tourism, increasing community capability is vital to manage tourism based on existing potential. In addition, people who are capable of managing tourism can also affect tourist satisfaction during a tour.

Capability building and increasing community participation to be involved in tourism development require awareness of stakeholders and the government's role by providing various training programs, capacity building, and empowerment of tourism conscious communities (Manyara & Jones 2007). Tourism must be driven by the community, where members of the community are responsible for controlling the tourism infrastructure and facilities available around it (Bonewati et al. 2017; Mayaka et al. 2018). Participation is able to transform people's passive attitudes into responsible and profitable attitudes, inspire entrepreneurial ventures, build partnerships and collaborations, encourage a spirit of cohesiveness, and rejuvenate relationships between communities, tourism destinations, and external stakeholders to increase the prospects for successful and sustainable development (Aref & Redzuan 2009; Razzaq et al. 2012).

Various studies show that the failure of tourism development is due to low levels of education, unawareness, and limited means of gathering information (Aref & Redzuan 2009; Kala & Bagri 2018). Furthermore, poor levels of education, inadequate capacity, unconsciousness, anxious nature of society, and reluctance to take part in the decision-making process are significant obstacles in remote areas of Indonesia (Bonewati et al. 2017). The tourism development authorities' attitude, limited financial resources, poor community capacity, and unavailability of time also hinder community participation in rural destination areas (Ahmeti 2013). Unconsciousness, perceptions of negative aspects of tourism, seasonality, lack of proper training, and entrepreneurial skills hinder local people's participation in tourism (Kala & Bagri 2018). Lack of awareness, general knowledge about tourism, local leadership, entrepreneurial skills, organizational structure, and networks are also some of the common obstacles to effective tourism development (Aref & Redzuan 2009; Razzaq et al. 2012). A large part of the basis for sustainable bottom-up community development lies in building community capacity. Therefore, community capacities or capabilities must be developed even before a community project or initiative is initiated. Tourism development must be planned in parallel with increasing community capacity (Dangi & Jamal 2016; Dodds et al. 2018; Mtapuri & Giampiccoli 2016). In addition, community capacity development emphasizes collaborative, sustainable, and influential processes based on human relations for the development process (Aref & Redzuan 2009; Kala & Bagri 2018; Razzaq et al. 2012; Zielinski et al. 2018).

4 CONCLUSION

The long debate regarding the concept of CBT and its implications for tourism development, in the end, theoretically and empirically indicate the importance of community knowledge and capabilities to be directly involved in developing tourist destinations. Community participation and the ability to implement their expertise in maintaining, developing, and innovating are the keys to sustainable tourism development. In addition, the role of stakeholders in accommodating and empowering local communities in destination development is another key concern for scholars.

REFERENCES

Ahmeti, F. 2013. Building Community Capacity for Tourism Development in Transitional Countries: Case of Cossovo. *European Journal of Scientific Research*, Vol. 115 No. 4, pp. 536–543.

Amilia, W., Rokhani, Prasetya, RC. and Suryadharma, B. 2020. Pembangunan Desa Wisata Gadingan dan Kebutuhan Pengembangan Sumber Daya Manusia dalam Pendekatan Community Based Tourism. *JPPM (Jurnal Pengabdian Dan Pemberdayaan Masyarakat)*, Vol. 4 No. 1, pp. 93–102.

Aref, F. and Redzuan, M.B. 2009. Community Capacity Building for Tourism Development. *Journal of Human Ecology*, Vol. 27 No. 1, pp. 21–25.

Bonewati, Salman, D. and Barkey, AR. 2017. Peningkatan Kapasitas Masyarakat dalam Pengembangan Pariwisata Berbasis Masyarakat di Desa Olele Kabupaten Bone Bolango. *Jurnal Analisis*, Vol. 6 No. 2, pp. 139–144.

Butler, RW. 1980. The concept of a tourist area cycle of evolution: Implications for management of resources. *The Canadian Geographer*, Vol. 24 No. 1, pp. 5–12.

Curcija, M., Breakey, N. and Driml, S. 2019. Development of a conflict management model as a tool for improved project outcomes in community based tourism. *Tourism Management*, Vol. 70, pp. 341–354.

Dangi, T. and Jamal, T. 2016. An Integrated Approach to 'Sustainable Community-Based Tourism'. *Sustainability*, Vol. 8 No. 5, p. 475.

Dodds, R., Ali, A. and Galaski, K. 2018. Mobilizing knowledge: determining key elements for success and pitfalls in developing community-based tourism. *Current Issues in Tourism*, Vol. 21 No. 13, pp. 1547–1568.

Ernawati, N.M., Sanders, D. and Dowling, R. 2017. Host-Guest Orientations of Community-based Tourism Products: A Case Study in Bali, Indonesia. *International Journal of Tourism Research*, Vol. 19 No. 3, pp. 367–382.

Inskeep, E. 1991. *Tourism Planning*, Van Nostrand Reinhold, New York.

Kala, D. and Bagri, S.C. 2018. Barriers to local community participation in tourism development: Evidence from mountainous state Uttarakhand, India. *Tourism: An International Interdisciplinary Journal*, Vol. 66 No. 3, p. 16.

Kim, S. and Kang, Y. 2020. Why do residents in an overtourism destination develop anti-tourist attitudes? An exploration of residents' experience through the lens of the community-based tourism. *Asia Pacific Journal of Tourism Research*, Vol. 25 No. 8, pp. 858–876.

Lawson. 1998. *Tourism and Recreation Handbook of Planning and Design*, Architectural Press, London.

Lee, T.H. and Jan, F.-H. 2019. Can community-based tourism contribute to sustainable development? Evidence from residents' perceptions of the sustainability. *Tourism Management*, Vol. 70, pp. 368–380.

Lee, T.H., Jan, F.-H. and Yang, C.-C. 2013. Conceptualizing and measuring environmentally responsible behaviors from the perspective of community-based tourists. *Tourism Management*, Vol. 36, pp. 454–468.

Lindstrom, K.N. and Larson, M. 2016. Community-based tourism in practice: evidence from three coastal communities in Bohuslän, Sweden. *Bulletin of Geography. Socio-Economic Series*, Vol. 33 No. 33, pp. 71–78.

Manyara, G. and Jones, E. 2007. Best practice model for community capacity-building: A case study of community-based tourism enterprises in Kenya. *Tourism: An International Interdisciplinary Journal*, Vol. 55 No. 4, p. 13.

Mayaka, M., Croy, W.G. and Cox, J.W. 2018. Participation as motif in community-based tourism: a practice perspective. *Journal of Sustainable Tourism*, Vol. 26 No. 3, pp. 416–432.

Mtapuri, O. and Giampiccoli, A. 2016. Towards a comprehensive model of community-based tourism development. *South African Geographical Journal*, Vol. 98 No. 1, pp. 154–168.

Ngo, T., Lohmann, G. and Hales, R. 2018. Collaborative marketing for the sustainable development of community-based tourism enterprises: voices from the field. *Journal of Sustainable Tourism*, Vol. 26 No. 8, pp. 1325–1343.

Noho, Y. 2014. Kapasitas Pengelolaan Desa Wisata Religius Bongo Kabupaten Gorontalo. *Jurnal Nasional Pariwisata*, Vol. 6 No. 1, pp. 8–21.

Okazaki, E. 2008. A Community-Based Tourism Model: Its Conception and Use. *Journal of Sustainable Tourism*, Vol. 16 No. 5, pp. 511–529.

Pemayun, A.A.G.P. and Maheswari, A.A.I.A. 2017. Economic Impacts of Craftsman Statue on Community Based Tourism Development. *International Journal of Social Sciences and Humanities (IJSSH)*, Vol. 1 No. 3, p. 59.

Purnomo, A., Idris, I. and Kurniawan, B. 2020. Understanding local community in managing sustainable tourism at Baluran National Park – Indonesia. *GeoJournal of Tourism and Geosites*, Vol. 29 No. 2, pp. 508–520.

Ramadhan, F. 2014. Partisipasi Masyarakat dalam Mendukung Kegiatan Pariwisata di Desa Wisata Bejiharjo, Gunungkidul, Yogyakarta. *Jurnal Teknik PWK*, Vol. 3 No. 4, pp. 949–963.

Razzaq, A.R.A., Mustafa, M.Z., Suradin, A., Hassan, R., Hamzah, A. and Khalifah, Z. 2012. Community Capacity Building for Sustainable Tourism Development. *Business and Management Review*, Vol. 2 No. 5, pp. 10–19.

Ruiz-Ballesteros, E. and Cáceres-Feria, R. 2016. Community-building and amenity migration in community-based tourism development. An approach from southwest Spain. *Tourism Management*, Vol. 54, pp. 513–523.

Sastropoetro, S. 1986. *Partisipasi, Komunitas, Persuasi, Dan Disiplin Dalam Pembangunan Nasional*, Alumni, Bandung.

Setyanto, I. 2019. Pengaruh Komponen Destinasi Wisata (4A) Terhadap Kepuasan Pengunjung Pantai Gemah Tulungagung. *Jurnal Administrasi Bisnis (JAB)*, Vol. 72 No. 1, pp. 157–167.

Sofield, T. 2003. *Empowerment for Sustainabel Turism Development*, Pergamon, Elsevier Science, Oxford.

Sulistiyorini, N.R. 2015. Partisipasi Masyarakat dalam Pengelolaan Sampah di Lingkungan Margaluyu Kelurahan Cicurug. *Share Social Work Jurnal*, Vol. 5 No. 1, pp. 71–80.

Utama, IG. 2017. *Pemasaran Pariwisata*, Andi, Yogyakarta.

Wahyuningtyas, N., Kodir, A., Idris, I. and Islam, N. 2020. Accelerating tourism development by community preparedness on disaster risk in Lombok, Indonesia. *GeoJournal of Tourism and Geosites*, Vol. 29 No. 2, pp. 545–553.

Wulaningrum, PD. 2018. Pengembangan Kawasan Wisata Berbasis Partisipasi Masyarakat Lokal di Koripan 1 Dlingo. *Jurnal BERDIKARI*, Vol. 6 No. 2, pp. 131–140.

Zielinski, S., Kim, S., Botero, C. and Yanes, A. 2018. Factors that facilitate and inhibit community-based tourism initiatives in developing countries. *Journal Current Issues in Tourism*, Vol. 23 No. 6, pp. 723–739.

Community Empowerment through Research, Innovation and Open Access – Sayono et al (Eds)

Outdoor learning based on natural laboratory as social studies learning resources for strengthening student's insights and characters

B. Kurniawan*, S.M. Towaf, Sukamto, A. Purnomo & Idris
Universitas Negeri Malang, Malang, Indonesia

ABSTRACT: Through innovative learning, students' insight and character are expected to be improved. Various efforts have been made by the teacher to achieve these goals, one of which is by using outdoor learning based on the natural laboratory. This study aims to explore the potential of learning resources in the natural laboratory managed by the Faculty of Social Sciences, State University of Malang, regarding social studies learning. This research was conducted in Gandusari District, Blitar-Indonesia. By using a qualitative descriptive approach, this study finds that historical potential has an impact on increasing students' insights regarding historical material in social studies learning. This research also shows that there is a strengthening of student character and meaningful social studies learning

Keywords: outdoor learning, natural laboratory, social studies, social insight, students' character

1 INTRODUCTION

Knowledge exploration can not only be carried out in a class since its situation is multidimensionally limited by time (Voss et al. 2017), with complicated class management (Schmidt et al. 2017) and learning partners such as classmates and teachers (Brantley 2014). This exploration can also be performed outside the classroom (McGowan 2016). Studying in class limits the experience obtained by the students. Springer and Collins (2008) consider formal classroom learning, unable to capture all components of the real-world situation, although this can create interactions between students. James & Williams (2017) add that classroom learning involving educators, textbooks, laboratory space, and discussion activities is not enough to develop the concept of understanding. Therefore, it is necessary to design learning that provides a more dynamic and meaningful experience.

One dynamic learning that provides experience is by directly sending students to the field, commonly referred to as outdoor learning (Williams et al. 2018). In the field, they interact with books, teachers, and friends as they usually do in class, as well as interact directly with learning resources. Learning resources can be objects (Bento & Dias 2017), the environment, and also society (McGowan 2016). In addition, they can also do observations, research, and experiment. They can observe and research a society's culture directly in the form of historical heritage, values, customs, and others. This is called the natural laboratory.

The Faculty of Social Sciences, Universitas Negeri Malang, has established Gadungan Village, Blitar Regency as its natural laboratory area since 2015. This village holds historical stories and historical sites in the form of temples, archway structures of the Gadungan site, worship altar, and village founder's tomb. The temple was named Wringin Branjang, one of the tombs of *Tumenggung* Dermo Kusumo–the founder of the village or also called Eyang Dermo, who was the soldier of Prince Diponegoro. Uniquely, the people around the village have high historical awareness.

---

*Corresponding author: bayu.kurniawan.fis@um.ac.id

DOI 10.1201/9781003189206-27

The Gandungan community is able to treat cultural products in the form of temples and tombs (Malikhah 2017).

The existence of a natural laboratory provides learning experiences and unique insights for students. By studying in an environment outside the classroom, there are at least six target abilities that will be obtained by students, including 1) communication skills, 2) leadership, 3) cooperation in groups, 4) ability to understand nature, 5) skills in the field, and 6) awareness toward the environment (Gookin 2006). This article explores the potential obtained from outdoor learning activities in the natural laboratory managed by the Faculty of Social Sciences, Universitas Negeri Malang, especially as a learning resource for social studies learning.

2 METHODS

This study used a qualitative descriptive approach that aims to understand a phenomenon (Moleong 2010), such as understanding the historical potential found in natural laboratories in Blitar, Indonesia. The first step was to conduct a field survey by identifying historical potentials at the research location. The research location was in Gandusari sub-district, Blitar Regency, Indonesia, with an area of 88. 23 km^2 (BPS Kab. Blitar 2014). Data were generally collected using observation, interviews, and documentation (Arikunto 2014). The researcher conducted a field survey of the research location, identified existing historical potential, and conducted interviews with the local government and site managers, as well as ten local residents. In the field, researchers also took several pictures as documentary evidence. The collected data were analyzed in several stages, namely data collection, data reduction, data presentation, and conclusion or verification (Miles et al. 2014). In addition, checking data used triangulation (Denkin & Lincoln 2000) and peer discussion.

3 RESULTS AND DISCUSSION

3.1 *Historical potential for Social Studies learning in a natural laboratory*

The historical remains found at the location of the study came from the time of the Majapahit Empire until the time of the Indonesian independence struggle. Its forms include temples, cult sites, historical monuments, and sacred tombs (See Table 1). These relics can be used as the learning resource for meaningful learning.

Historical relics found at the research sites are generally influenced by the Majapahit Kingdom. This is useful to increase students' insight in learning ancient history. In addition to knowing the Majapahit Kingdom's greatness, students experience a natural process of character formation (Gookin 2006). In the field, students interact with learning resources, friends, and the surrounding community. The surrounding community can be the object of study in the fields of history, social, and culture. The reason is that the village community highly values cultural values and safeguards

Table 1. Historical site in Gandusari sub-district, Blitar.

Name	Description
Kotes Temple	Temples/Relics of Majapahit Emipre
Wringin Brajang Temple	Temples /Relics of Majapahit Emipre
Sumberagung Temple	Temples /Relics of Majapahit Emipre
Gunung Gedhang Site	Cult sites/Relics of Majapahit Emipre
Sukosewu Site	Cult sites/Relics of Majapahit Emipre
Slumbung Site	Cult sites/Relics of Majapahit Emipre
Monumen TRIP	Historic monument/Era of Indonesian independence strunggle
Tomb of *Eyang* Darmo	Sacred tombs/Gadungan Village founder and soldier of Prince Diponegoro

these historical sites. The temple has become an authentic medium often used by educators for learning related to history, for example, Borobudur Temple, Prambanan Temple, Penantaran Temple, and others. The research conducted by Permadi (2015) on Singhasari Temple as a learning tool for students with tourism methods. The temple's tourism method can provide awareness in preserving historical relics for teachers, students, communities, and local governments. In addition, the temple can also be used as an inspiration for learning based on computer media. Research conducted by Hidayat et al. (2014) presents material about parts of the Borobudur Temple into a video game. As a result, students get knowledge related to Borobudur Temple and fun learning packaged in a game.

Temples and other historical sites as social studies learning objects can be studied in an integrated manner. For example, there are several things that can be studied in terms of historical, socio-cultural, and economic aspects. In terms of history, students can directly examine the establishment of temples, patterns, and functions of the temple. In the socio-cultural aspect, students can study the community and its social activities, the values contained in the temple, and its people. In addition, in terms of economics, students can examine whether the temple has an economic impact or not for the community, for example, by presenting tourism and others.

Outdoor learning, recreation, and outdoor programs using historical resource sites as described earlier provide experience to help facilitate various student personal growth types such as individual and group social systems, independence, and resilience (Chang et al. 2019). In addition, this learning model positively affects individual psychological well-being by offering arrangements that can reduce stress (D'Amato & Krasny 2011). In fact, in certain situations, this model can be an effective way to restore their psychosocial (Mutz & Müller 2016). Overall, much of the literature proves that outdoor learning-based activities can have a positive effect on the whole individual, especially during their academic careers (Chang et al. 2019).

3.2 *Character-based social studies learning*

Based on the findings, knowledge source facilities are very supportive in implementing integrated social studies learning. Geological objects can be viewed historically, sociologically, and economically as interesting objects to study. In addition, character-building is an essential element to be achieved.

Students who are invited to study the historical aspects, especially the historical temple sites, and the sacred tomb of *Eyang* Darmo understand some of the meanings and values contained. At least there are several characters that can be obtained, such as 1) religiosity, 2) tolerance, 3) curiosity, 4) national spirit, and 5) caring for the environment. The TRIP monument can also shape students' character, such as love for the country or nationalism and hard work. In terms of sociology, students can learn how relationships between community members can live in harmony and peace and how social construction is woven to always safeguard every cultural asset in their environment. These characters are expected by every student through learning in the natural laboratory.

Learning social studies on an outdoor basis offers students the opportunity to develop knowledge and skills by expanding their daily experience in the classroom against modern life's negative effects. (Romar et al. 2019). Through direct interaction with the community and looking deeply into the realities of people's lives, students unconsciously build a system of values and characters in themselves through experiences that are absorbed through the realities of life that arise in social life (Kurniawan 2016; Mpeli & Botma 2015; Thorburn 2018).

3.3 *Meaningful social studies learning*

Students need a different and fresher learning experience. Knowledge gained is based on knowledge transformed by the teacher based on textbooks. James dan Williams (2017) explains that students get good knowledge when learning is active, provides meaning and experience. The application of learning in natural laboratories in Blitar Regency provides new knowledge directly from the place, constructivist learning activities, and meaningful experiences.

Students need an understanding of attitude values to build good character. This means that learning requires value clarification techniques (VCT). Raths et al. (1966) described that there are 7 steps in using the value clarification approach, including choosing the value 1) freely, 2) from several alternatives, 3) after pondering it, 4) value that has been selected, 5) confirming to the public about the selected value, 6) take action according to the selected value, and 7) repeating or familiarizing the chosen value. Many studies apply the value clarification approach in learning, such as Guinn (1977) applies a value clarification approach in learning with students in cultural differences in America. Besides, Sariyatun (2011) applies value clarification techniques (VCT) in value-based social studies on cultural values contained in classical local batik. Meanwhile, in the field of health education, a value clarification approach is used to equip midwifery students. In addition, Mpeli & Botma (2015) apply a value clarification approach to providing a view for midwifery students related to abortion services in South Africa. This research is intended to answer the phenomenon of the high mortality rate of women due to abortion.

Regarding social studies learning, especially the historical heritage in Gandusari, students interact with communities around the village with high historical awareness. Students need value clarification techniques to understand and internalize cultural values. Students can also build an understanding of historical awareness by knowing the meanings contained in historical sites. Therefore, if there has been historical awareness in students, there will be no human activities that damage historical sites or even destroy them. Destroying historical sites shows no respect toward the inheritance of earlier ancestors. On the other hand, not necessarily the people today are able to create what have been made by previous ancestors. One of the factors destroying historical sites is human advancement. The newspaper "Monitor Depok," which was published on Friday, June 23, 2006, in column 6, page 1, reported a historic old house's demolition. The ancient house belongs to an ethnic Chinese named Tan. This happens because there is a lack of understanding and awareness of heritage and historical sites' importance. Thus, historical sites will be increasingly marginalized by developments, especially those in big cities.

The natural laboratory may often be heard in the learning of the natural sciences. Whereas, in learning, social sciences are known as field practices (Muralidharan 2017), learning outside the classroom, or outdoor learning (James & Williams 2017). Related to social studies learning, natural laboratories can provide fresher experiences to students. They not only interact directly with the source of knowledge (Bento & Dias 2017) but also they can shape leadership attitudes (Sani et al. 2018), learn to take risks, build supportive relationships and interact with the community (Daniel et al. 2014). This will be a good thing for students' character development. In addition, by providing experiences in the field, students can integrate their experiences in their writing (notes) and analysis (White 2016). It also expands student understanding of the social economy of the society (Kurniawan 2017) and encourages them to explore the meaning of any symbols existing in social heritage (Kurniawan 2016). Thus, outdoor learning provides a complete package of nutrients needed by students to develop into good individuals and citizens.

Most people think that learning in a natural laboratory is full of risks. This is based on the reason that learning outside the classroom can increase physical risk and expand the potential for psychological anxiety (Canter et al. 1994). However, some previous literature indicates that anxiety conditions, external disorders, and risk situations can encourage individuals to create ways or solutions to overcome them cognitively (Luckner & Nadler 1997). This is supported by several research results that indicate a positive impact on the program (Outdoor Learning) to increase self-confidence, good communication, mutual help, and dare to express opinions (Neill 2003). There are also benefits for children's health and development in the form of motor development and physical health (Bento & Dias 2017). In the natural laboratory, students can communicate directly to the surrounding community and collect data for learning purposes. This can help a lot in historical site studies where students need to see, touch, and understand the object directly (Yesilbursa & Barton 2011), to get significant results in learning (Blair 2016). They will reconstruct the knowledge and get the unforgettable meaning of learning.

4 CONCLUSION

Research on natural laboratory-based outdoor learning provides a different learning atmosphere that presents valuable experiences for students. Their insight is increasing because they learn directly from learning resources in the field. The results show that the natural laboratory managed by the Faculty of Social Sciences, Universitas Negeri Malang, has excellent historical potential for social studies learning related to historical material. This historical potential has an impact on improving the character of students in the form of 1) religiosity, 2) tolerance, 3) curiosity, 4) environmental care, 5) nationalism, 6) hard work, and 7) caring for local culture. In addition, learning directly from the learning resources provides students with meaningful social studies learning experiences.

REFERENCES

Arikunto, S. 2014. *Prosedur Penelitian: Suatu Pendekatan Praktik*. Jakarta: Rineka Cipta.

Autry, C. E. 2001. Adventure therapy with girls at-risk: Responses to outdoor experiential activities. *Therapeutic Recreation Journal*, *35*(4), 289–306.

Bento, G., & Dias, G. 2017. The importance of outdoor play for young children's healthy development. *Porto Biomedical Journal*, *2*(5), 157–160.

Blair, D. j. 2016. Experiental learning for teacher professional development at historical sites. *Journal of Experiental Education*, *39*(2), 1–15.

BPS Kab. Blitar. 2014. *Statistik Daerah Kecamatan Gandusari 2014*. Blitar: BPS Kab. Blitar.

Brantley, A. 2014. Learning Outside the Classroom. *Phil Delta Kappan*, *98*(8), 70–75.

Canter, M. B., Bennett, B. E., Jones, S. E., & Nagy, T. F. 1994. *Ethics for psychologists: A commentary on the APA Ethics Code*. Washington DC: American Psychological Association.

Chang, Y., Davidson, C., Conklin, S., & Ewert, A. 2019. The impact of short-term adventure-based outdoor programs on college students' stress reduction. *Journal of Adventure Education and Outdoor Learning*, *19*(1), 67–83. https://doi.org/10.1080/14729679.2018.1507831

D'Amato, L. G., & Krasny, M. E. 2011. Outdoor adventure education: Applying transformative learning theory to understanding instrumental learning and personal growth in environmental education. *The Journal of Environmental Education*, *42*(4), 237–254.

Daniel, B., Bobilya, AJ., Kalisch, KR., & McAvoy, LH. 2014. Autonomous student experiences in outdoor and adventure education. *Journal of Experiential Education*, *37*(1), 4–17.

Denkin, N. K., & Lincoln, Y. S. 2000. *Handbook of qualitative research*. US.: Thousand Qaks.

Gookin, J. 2006. *NOLS Wilderness Educator Notebook*. WY: National Outdoor Leadership School.

Guinn, R. 1977. Value clarification in the bicultural classroom. *Journal of Teacher Education*, *28*(1), 46–47.

Hidayat, IK., Sunarto, P., & Guntur. 2014. Mengenal relief, mudra, dan stupa candi borobudur untuk anak-anak usia 9-12 tahun melalui Edugame. *Jurnal Visual, Art & Design*, *6*(1), 58–68. https://doi.org/doi:10.5614/itbj.vad.2014.6.1.6.

James, J. K., & Williams, T. 2017. School-Based Experiental Outdoor Education: A Neglected Necessity. *Journal of Experiental Education*, *40*(1), 58–71.

Kurniawan, B. 2016. *Model pembelajaran IPS berbasis nilai simbolisme kain Songket untuk meningkatkan solidaritas sosial siswa di SMP Negeri 6 Kayuagung* (PhD Thesis). Universitas Sebelas Maret.

Kurniawan, Bayu. 2017. Menumbuhkan jiwa wirausaha dalam pembelajaran IPS (Belajar dari etos kerja dan berdagang masyarakat Minangkabau). *Prosiding Seminar Nasional PIPS*, 56–62. Malang: UM Press.

Luckner, J. L., & Nadler, R. S. 1997. *Processing the experience: Strategies to enhance and generalize learning*. Dubuque: Kendall/Hunt Publishing Company.

Malikhah, S. 2017. *Pengembangan pembelajaran IPS tematik terpadu berbasis laboratorium alam Fakultas Ilmu Sosial*. Malang: UM Press.

McGowan, A. L. 2016. Impact of one-semester outdoor education programs on adolescent perceptions of self-authorship. *Journal of Experiential Education*, *39*(4), 386–411.

Miles, M. B., Huberman, A. M., & Saldana, J. 2014. *Qualitatif Data Analysis A Methods Sourcebook Edition 3* (Terjemahan Tjetjep Rohindi Rohidi, UI Press). USA: Sage Publication.

Moleong, LJ. 2010. *Metode Penelitian Kualitatif*. Bandung: Remaja Rosdakarya.

Monitor Depok. 23 Juni 2006. *Ekses Pembangunan Situs Sejarah di Depok Musnah*, hlm.1.

Mpeli, M. R., & Botma, Y. 2015. Abortion-related services: Value clarification through 'difficult dialogues' strategies. *Journal of Education, Citizenship, and Social Justice, 10*(3). Retrieved from http://journals. sagepub.com/doi/full/10.1177/1746197915607281

Muralidharan, K. 2017. Field experiments in education in developing countries. *Handbook of Economic Field Experiments, 2*, 323–385.

Mutz, M., & Müller, J. 2016. Mental health benefits of outdoor adventures: Results from two pilot studies. *Journal of Adolescence, 49*, 105–114.

Neill, J. T. 2003. Reviewing and benchmarking adventure therapy outcomes: Applications of meta-analysis. *Journal of Experiential Education, 25*(3), 316–321.

Permadi, GS. 2015. *Pemanfaatan candi Singhasari sebagai sumber belajar sejarah.* Jember: Universitas Jember.

Raths, L. E., Harmon, M., & Simon, S. B. 1966. *Values and teaching.* Columbus, Ohio: Charles E. Merrill Books, Inc.

Romar, J.-E., Enqvist, I., Kulmala, J., Kallio, J., & Tammelin, T. 2019. Physical activity and sedentary behaviour during outdoor learning and traditional indoor school days among Finnish primary school students. *Journal of Adventure Education and Outdoor Learning, 19*(1), 28–42. https://doi.org/10.1080/ 14729679.2018.1488594

Sani, A., Ekowati, VM., Wekke, I. S., & Idris, I. 2018. Respective contribution of entrepreneurial leadership through organizational citizenship behavior in creating employees performance. *Academy of Entrepreneurship Journal, 24*(4), 1–11.

Sariyatun. 2011. Social studies learning model in local classic batik cultural based for the nation identity reinforcement of junior high school students. *International Journal of History Education, 12*(2).

Schmidt, J., Klusmann, U., Lüdtke, O., Möller, J., & Kunter, M. 2017. What makes good and bad days for beginning teachers? A diary study on daily uplifts and hassles. *Contemporary Educational Psychology, 48*, 85–97. https://doi.org/doi:10.1016/j.cedpsych.2016.09.004

Springer, S., & Collins, L. 2008. Interacting inside and outside of the language classroom. *Language Teaching Research, 12*(1), 39–60.

Thorburn, M. 2018. Moral deliberation and environmental awareness: Reviewing Deweyan-informed possibilities for contemporary outdoor learning. *Journal of Adventure Education and Outdoor Learning, 18*(1), 26–35. https://doi.org/10.1080/14729679.2017.1322000

Voss, T., Wagner, W., Klusmann, U., Trautwein, U., & Kunter, M. 2017. Changes in beginning teachers' classroom management knowledge and emotional exhaustion during the induction phase. *Contemporary Educational Psychology, 51*, 170–184. https://doi.org/doi:10.1016/j.cedpsych.2017.08.002

White, J. 2016. historical sociology in the field: Teaching irish identity through field experience. *Irish Journal of Sociology, 24*(1), 54–77.

Williams, I. R., Rose, L. M., Olsson, C. A., Patton, G. C., & Allen, N. B. 2018. The impact of outdoor youth programs on positive adolescent development: Study protocol for a controlled crossover trial. *International Journal of Educational Research, 87*, 22–35. https://doi.org/doi:10.1016/j.ijer.2017.10.004

Yesilbursa, C. C., & Barton, K. C. 2011. Preservice teachers attitudes toward the inclusion of "heritage education" in elementary social studies. *Journal of Social Education Research, 2*(2), 1–21.

Community Empowerment through Research, Innovation and Open Access – Sayono et al (Eds)
© 2021 Copyright the Author(s), ISBN 978-1-032-03819-3

Reorganizing the *Ummah*: COVID-19 and social transformation in plural society

A.A. Widianto*, L.A. Perguna, T. Thoriquttyas & F. Hasanah
Universitas Negeri Malang, Malang, Indonesia

ABSTRACT: COVID-19, which has hit various countries in the world, has brought significant changes to social life. In this context, the pandemic has also contributed to social transformation in the plural society in Sukoreno. Sukoreno village is known as the village of Pancasila because of its religious pluralism. This study aims to describe the impacts of a pandemic on a plural society and ways to respond to it. This research uses qualitative methods with data collection through interviews and observations. Interviews were conducted with village officials and religious leaders. The data analysis technique used was an interactive model. The results of this study indicate that the COVID-19 pandemic has led to socio-religious transformation. Various policies, regulations and agreements related to handling COVID-19 changed socio-religious routines and interfaith social relations. Routines and annual religious activities should be modified according to health policies and protocols. On the other hand, there has been a strengthening of interfaith relations in dealing with the pandemic's impact. Interfaith collaboration or community gathering forums (SILAMAS) are formed by optimizing their respective resources to help people who need assistance. This social reality further emphasizes that the pandemic, on the one hand, results in social transformation and affects social relationships between elements of society.

Keywords: *Ummah*, plural society, *Desa Pancasila*

1 INTRODUCTION

The world is facing a COVID-19 catastrophic outbreak, which until now its spread is increasing in several countries, including Indonesia. In Indonesia, there has been a rapid increase of people infected with COVID-19. From only 2 positive cases announced on March 2, 2020, on May 20, it has reached 18,496 positives, 4,467 recovered, and 1,221 deaths (World Health Organization 2020). In practice, various legal products and policies for handling COVID-19 are not effective in suppressing the escalation of the number of cases in Indonesia (Djalante et al. 2020). One of the causes is the policy crisis in handling COVID-19 (Widianingrum & Mas'udi 2020; Hidayat 2020). In addition, the policy approach tends to be top-down and without optimizing the potential of local culture and institutions. In fact, various levels and community institutions are very prospective in helping to handle COVID-19 (Djalante et al. 2020).

The increasing spread of COVID-19 in Indonesia is a serious problem in the medical field and in other strategic fields at various levels. In the socio-religious aspect of society, COVID-19 has had an impact on changing patterns of worship due to restrictions on mass worship activities (Post 2020). Even though the restrictions were applied, some communities violated them, creating new distribution clusters (Quadri 2020). On the other hand, religion can play a productive role as well.

The impact of COVID-19 also targets the people of Jember as a melting pot area where various ethnic groups mingle and coalesce, especially in the Sukoreno village Jember, which is as popular as

---

*Corresponding author: ahmad.arif.fis@um.ac.id

DOI 10.1201/9781003189206-28

Pancasila Village because it can build religious harmony. Religious adherents in Sukoreno consist of 7152 Muslims, 35 Protestants, 110 Catholics, 124 Hindus and 15 Buddhists (BPS 2018) as well as followers of the *Sapta Darma* and *Ilmu Sejati* beliefs. Apart from its plurality, Sukoreno Village was chosen because its people can maintain social cohesion and harmony amidst diversity. The tolerance of the Christian-Muslim community in Jember is also driven by ethnic and cultural similarities and the equality of socio-economic classes (Mamuaya & Sair 2017). Post-religious conflict resolution is also built on local wisdom and obedience to religious leaders (Putra 2013).

Several studies show that many interfaith institutions have not been optimized even though they have contributed to the handling of COVID-19, such as the Indonesian Ulema Council (MUI), the Indonesian Church Association (PGI), the Indonesian Buddhist Council, and so forth (Djalante et al. 2020). In the context of a plural society, religious diversity can be a potential resource in dealing with COVID-19. However, most of them present a limited contribution. In fact, these resources will certainly be more productive if they are engaged in interfaith cooperation relations. Therefore, it is crucial to see the impact of the pandemic on socio-religious life and the response of the interfaith community in handling COVID-19 in plural societies. This study aims to describe the local community's response across religions against the pandemic and map the social-religious changes.

2 METHODS

This research was conducted in Sukoreno Village, Umbulsari District, Jember Regency. The focus of this research was geographic and sociological, which described the complexity of community relations in the social spaces within it. The subject of this research was the Sukoreno community, which consisted of plural social elements, especially multi-religions and religious sects. This research used qualitative methods to explore data and an in-depth understanding of social phenomena (Silverman 2005)

Data collection was carried out through observation with the aim of (1) understanding social settings naturally, (2) capturing events that affected social processes and the focus of research, (3) identifying the regularity or repetition of social realities in society, and (4) understanding social reality from a research subjects' perspective (Black & Champion 2009). Observations were carried out in a participatory manner in the Sukoreno community, focusing on observing social processes, social institutions and structures, patterns of interaction and relationships between religious groups, ongoing traditions, symbols and social realities that influence the focus of this research. To enrich the data, interviews were conducted with informants to obtain a comprehensive picture of the required data. Informants were selected based on the purposive informant selection technique, based on the ability and accuracy of the informant regarding the required data.

3 RESULTS AND DISCUSSION

3.1 *Socio-historical context of plural society in Sukoreno, Jember*

Jember, which is the focus of this research, is prone to religious conflict, but also has positive potential in building social harmony and cohesion as in the Sukoreno village. Quoting Casanova (in Sumrahadi 2018), in religion, two identities are attached at once. One side of religion is exclusive, particular and primordial. On the other hand, religion displays inclusiveness and transcendentalism as well. Religion can act as social glue or break the existing harmonies between religious communities (Nottingham 1993, Thomas O'dea 1987).

In the context of Jember, several studies discuss religious pluralism and religious-based conflicts. These studies discuss: (1) the ethics of the *Pandalungan* community in creating diversity (religion) (Sair 2019); (2) tolerance of the Madura Muslim-Christian community in the Sumberpakem Village (Mamuaya & Abdus Sair 2017); (3) harmonization of diversity and interfaith dialogue in Sukoreno

Jember (Rosadi 2018); (4) the "Sunni-Shia" religious conflict in the Puger District (Izzati 2018); (5) the role of local wisdom in the resolution of religious belief conflicts in East Java (Putra 2013); and (6) communal consolidation as mitigation of religious conflict in East Java (Fattah 2018).

Jember is an area that has cultural and religious diversity. This can be oriented towards the extent of existing social and religious conflicts of Shia and Sunni conflict in the Puger District, Jember (Ghufron 2015). In addition, there are also counter-arguments that occur in the Puger community regarding the assumption of taking *Isut* in Islamic law (Qomariyah & Sholihin 2019). Various roots of conflict in the Jember area can be attributed to the existing culture in the region. *Pendhalungan* is a cultural aspect is an effort to assimilate tribes and their differences (Sasmita & Endang Widuatie 2016). Besides various conflicts that exist in the Jember region, strengthening the aspect of tolerance in Sukoreno Village plays a role in modeling for other areas experiencing religious and cultural disintegration.

According to data on religious diversity in Jember BPS Provinsi Jawa Timur (2016), it is recorded that Jember contained 2.294.519 Muslims, 28.926 Protestants, 19.288 Catholics, 1.609 Hindus, 3.401 Buddhists and 343 other believers. Multiculturalism is an ideology that can achieve a prosperous community life (Suparlan 2014). Sukoreno Village, which has been lined up to become the Pancasila village in the Jember Regency, demonstrates religious harmony by the residents harmonizing with each other and holding religious ceremonies or celebration of the holidays of each of the existing religions.

Discussions about religion have never been questioned by residents. This is due to the high tolerance and understanding of individuals and religious groups to respect each other religions. The main aspect that encourages social cohesion among them is the difference in religious differences that already existed and were brought by ancestors so that the issue of identity never seems to gain traction. However, that does not mean that there have never been conflicts between religions.

3.2 *Pandemic and social transformation in Sukoreno*

In following up on local government appeals and regulations in implementing policies to anticipate and prevent the spread of the Coronavirus, on March 26, 2020, body temperature checks were held when Hindus were about to perform worship at the temple in the *Ngembak Geni* ceremony at Swasty Dharma Temple in Sukoreno Village, Umbulsari, Jember. There are also rules for washing hands and using hand sanitizers provided by local health center officers who joined as a health unit in this Hindu event. Hindus who entered the temple were told they must also go through a disinfectant spray.

The formulation of policies in Sukoreno village certainly comes from above, where the district government stipulates that each village must meet health protocols in carrying out all existing activities. In addition, the village of Sukoreno itself has its health protocol at each place of worship. The COVID policy's termination in Sukoreno village was approved by the local village head and decided as an index of health protocols in the Suskoreno area.

In handling COVID-19 in Sukoreno, the interfaith community also played an active role in handling the COVID task force, COVID volunteers and youth organizations. The involvement of interfaith community leaders is based on the principle of cooperation and harmony among fellow citizens. The local people never question each other's religion in this matter. The enthusiasm of the interfaith community was manifested when carrying out tightening health protocols at guard posts between regions, as well as during religious activities, wherein this case it was found that when carrying out Eid prayers in 2020, which was in the midst of the pandemic the previous day, parking guards and health workers maintained order within the interfaith community. This was intended to ensure conducive conditions in the mosque, and vice versa. As for the COVID formation mechanism, it was initiated by the local village government and youth organizations by conducting deliberations with religious leaders and then religious leaders channeling it to congregations, so that the formation of an interfaith COVID volunteer team was formed to secure the Sukoreno village.

3.3 The power of cross-religion solidarity in managing the pandemic

As explained above, in preventing the spread of COVID, all Sukoreno residents joined in handling COVID through the village COVID task force team, COVID volunteers, and youth organizations. Each of these communities cooperates with each other in Sukoreno village. Each community function is divided based on its work assignments. The assignments consist of spraying regularly at every resident's house, maintaining monitoring posts, collecting data on vulnerable people based on age and immune systems due to illness, providing social assistance to poor people, especially widows and maintaining order when carrying out religious activities. In their respective duties, the community can contribute to each other to maintain the village's security and its people, so that the problem of practicing religion is sidelined to realize the social activities that are being carried out.

All this COVID collaboration has also led to deeper intimacy between communities. The community does not depend on the COVID handling community alone, but all local people are harmonious to maintain and obey the existing health protocols in Sukoreno village. The collaboration results in no people identified as being infected by the Coronavirus, so the village or area is categorized in the yellow zone.

In anticipating COVID-19 in Sukoreno, there are many rituals and religious ceremonies that are independent of the existing village government policies. Rituals and religious ceremonies are carried out by each religion with the religious leaders who guide the running of these rituals. This form of ritual is allowed by the village government to follow up on the prevention of COVID-19, which can disturb the local community. The people of Sukoreno also still believe that COVID-19 is a "*pagebluk*" which is a deadly disease without illness, so anticipation also comes from the local community by performing rituals in their respective religions. Of course, this ritual has received permission from the local village.

4 CONCLUSION

The COVID-19 pandemic has resulted in social changes in the people of Sukoreno Jember. The socio-religious life was reorganized in accordance with national policies and community consensus. Annual activities involving many parties must be modified to prevent the spread of the virus. One of the most affected aspects is the existence of restrictions on religious activities that cause crowds. Local-scale social restrictions reduce religious activity. However, few people still continue to carrying out congregational worship activities while still implementing health protocols. The social restriction policy during COVID-19 indirectly helped strengthen interfaith social relations in dealing with the impact of the pandemic. The leaders organize the resources they have in each congregation to work together to help restore the social conditions of the people affected by the pandemic. Through this collaboration, social relations become stronger and more productive for the community. This shows that in addition to changing socio-religious life, the pandemic has also affected inter-religious social ties based on local wisdom.

REFERENCES

Alfandi, M. 2013. Prasangka: Potensi Pemicu Konflik Internal Umat Islam. *Jurnal Walisongo*, Volume 21, No. 1. 23–41.

Berman, Yitzhak.2003. *Indikators For Social Cohesion*. Austria:The European Centre For Social Welfare Policy and Research.

BPS *Kabupaten* Jember. 2018. *Kecamatan Umbulsari Dalam Angka 2018*. Jember: BPS Jember.

Djalante, R., Lassa, J., Setiamarga, D., Sudjatma, A., Indrawan, M., Haryanto, B., Mahfud, C., Sinapoy, M. S., Djalante, S., Rafliana, I., Gunawan, L. A., Surtiari, G. A. K., & Warsilah, H. 2020. Review and analysis of current responses to COVID-19 in Indonesia: Period of January to March 2020.

Faisal, *Sanapiah*. 2015. *Pengumpulan Dan Analisis Data Dalam Penelitian Kualitatif*. Dalam *Burhan* Bungin. *Analisis Data Kualitatif*. Jakarta: Grafindo Persada.

Hamdi, Ahmad Zainul. 2013. Radicalizing Indonesian Moderate Islam From Within the NU-FPI Relationship in Bangkalan, *Madura Journal of Indonesian Islam.*Volume 07, No 01. 46–61.

Hidayat, R. 2020. *Teledor Penanganan Wabah COVID-19 di Indonesia.* tirto.id. Retrieved May 20, 2020, from https://tirto.id/teledor-penanganan-wabah-COVID-19-di-indonesia-eDPG

Ihsan Ali-Fauzi Rudy Harisyah Alam Samsu Rizal Panggabean. 2009. *Pola-Pola Konflik Keagamaan Di Indonesia (1990–2008).* Jakarta Kerjasama Yayasan Wakaf Paramadina (YWP) Magister Perdamaian Dan Resolusi Konflik, Universitas Gadjah Mada (MPRK-UGM), the Asia Foundation (TAF).

Izzati, Arini Robbi. 2018. Konflik Agama Antara "Sunni-Syiah" Di Kecamatan Puger, Kabupaten Jember. Dalam Optimalisasi Peran FKUB Mewujudkan Indonesia Damai. Penyunting Eko Riyadi & Despan Heryansyah. *Kalam: Jurnal Studi Agama Dan Pemikiran Islam*, Vol. 7, No. 1. 67–82.

Luthfi Assyaukanie 2018. *Akar-Akar Legal Intoleransi Dan Diskriminasi di Indonesia.*

Maarif, Ahmad Syafii 2012. *Politik Identitas Dan Masa Depan Pluralisme Indonesia* Dalam *Ihsan* Ali-Fauzi Dan Samsu Rizal Panggabean. Politik Identitas Dan Masa Depan Pluralisme Kita. Jakarta: Democracy Project Yayasan Abad Demokrasi.

Mahli Zainudin Tago. 2013. *Agama Dan Integrasi Sosial Dalam Pemikiran Clifford Geertz.*

Mamuaya, Chriestine Lucia & Abdus Sair. 2017. Toleransi Masyarakat Islam-Kristen Madura Di Desa Sumberpakem, Kecamatan Sumberjambe, Kabupaten Jember. *Dimensi.* Vol. 10. No.2, 76–86.

Mappiase, Sulaiman & Muliadi Nur. 2007. *Investigasi Konsep Pluralisme Keagamaan dan Loyalitas Masyarakat kepada Tokoh Agama di Sulawesi Utara.* Laporan Hasil Penelitian Kompetitif Kolektif Terpadu 2007.

Mohammad Takdir 2017. Identifikasi Pola-Pola Konflik Agama Dan Sosial (Studi Kasus Kekerasan Berbasis Sektarian Dan Komunal di Indonesia). *Jurnal RI'AYAH*, Vol. 02, No. 01. 15–32.

Muqoyyidin, Andik Wahyudin 2012. Potret Konflik Bernuansa Agama di Indonesia (Signifikansi Model Resolusi Berbasis Teologi Transformatif). *Analisis*, Volume XII, Nomor 2. 60–74.

Nottingham, Elizabet K. 1993. *Agama Dan Masyarakat: Suatu Pengantar Sosiologi.* Jakarta: Rajawali Press

Post, T. J. 2020. *Religion and COVID-19 mitigation.* The Jakarta Post. Retrieved May 20, 2020, from https://www.thejakartapost.com/academia/2020/03/26/religion-and-COVID-19-mitigation.html

Progress in Disaster Science, 6, 100091. https://doi.org/10.1016/j.pdisas.2020.100091

Quadri, S. A. 2020. COVID-19 and religious congregations: Implications for spread of novel pathogens. *International Journal of Infectious Diseases*, 96, 219–221. https://doi.org/10.1016/j.ijid.2020.05.007

Reychler, Luc. 2006. *Challenges of Peace Research. International Journal of Peace Studies*, Volume 11, Number 1, Spring/Sumer, 2006.

Ridwan, MK. Adang Kuswaya, Muhammad Misbah. 2016. *Agama; Antara Cita Dan Kritik.*

Rosadi Br. 2018. The Harmonization of Diversity and Interreligious Dialogue in Sukoreno Village Jember. *International Journal of Management and Administrative Sciences (IJMAS)*, Vol. 6, No. 2, (11–19).

Sair, Abdus 2019. Etika Masyarakat Pandalungan Dalam Merajut Kebhinekaan (Agama). *Jurnal Sosiologi Pendidikan Humanis JSPH* Vol. 4. No 1. 12–32.

Community Empowerment through Research, Innovation and Open Access – Sayono et al (Eds)
© 2021 Copyright the Author(s), ISBN 978-1-032-03819-3

Environmental ethics on Mount Kelud slopes: Investigating local community responses on the utilization of Mount Kelud materials

A.S.M. Fajar*
Institut Agama Islam Tribakti Lirboyo, Kediri, Indonesia

A.I. Mawardi
Universitas Islam Negreri Sunan Ampel, Surabaya, Indonesia

A. Syakur
STKIP PGRI Sidoarjo, Indonesia

ABSTRACT: Anchored by an ethnoecological approach, this study attempted to portray the contradictory situations from both the perspective of natural resources, the use of nature within the framework of an environmental ethics approach, and the cultural cognition that produces the patterns of adaptation from communities around Mount Kelud. Data in this study was gathered from the views of the community around the Kelud slope, which consists of social, religious, cultural and non-governmental organizations. Findings suggest that the existence and meaning of mythology and rituals are in the forms of religiously and locally based friendship life cycles. This study adds to the discussion of environmental ethics and the utilization of Mount Kelud materials in Kediri, East Java, Indonesia.

Keywords: environmental ethics, Kelud eruption, land resource

1 INTRODUCTION

Indonesia is geographically located along the Pacific Ring of Fire, causing the region to be dominated by active volcanic clusters, making Indonesia vulnerable to volcanic hazards. The presence of volcanoes can be a source of catastrophe that destroys life, especially from the eruptions and the lava. Volcanic eruptions often leave materials from the mountain that can damage plantations, agriculture, settlements, casualties, and other properties (Kelman & Mather 2008). However, a volcano is a phenomenal landscape that contains ambiguity because, besides being a source of, it is also a source of grace that is beneficial to human life around it, including soil fertility as agricultural and plantation land, a wealth of beneficial sand and stone materials, and the charm of nature with the beauty of the breathtaking mountain panorama (Tjandra 2015). Kelud volcano, which is located between three regencies in East Java province, namely Kediri, Blitar and Malang, is an active volcano that erupts within an interval of about 9–25 years. The explosion tends to be large and quick. The original character of the Kelud volcano is to have a crater lake that can produce a huge amount of lava flows when it erupts, and it is harmful to the surrounding population (Arsana 2014).

The eruption in 2014 showed that the Kelud volcano erupted explosively and in a short time within three hours. The rumbling and the booming sound could be heard as far as the other provinces, and the elevation of the volcanic column was 17 km, thus causing ash rain in the area of Central

---

*Corresponding author: bbssfwn@gmail.com

DOI 10.1201/9781003189206-29

Java and even some areas in West Java. The volcanic ash was even able to paralyze some airports in Java and flight disruption (Yusron 2018). The phenomenon is very different from the Kelud volcano eruption in 2007, beginning with an increase in the seismicity of September 2007. The eruption ended with forming a lava dome in the crater lake on November 3, 2007. This incident was a transition of previously explosive eruption properties to be effusive. On the observation with a GPS method for three sessions in April, August, and October 2008, it was seen that the vector pattern shifting each measuring point is more dominated and influenced by forces that work due to geological structure activities, especially the crustal stabilization phase after the final phase of the eruption in November 2007. The shifts caused by the magma flows are minor and occur only in measuring points that are near the Lava dome with a relatively shallow depth and associated with a seismic zone (Haerani et al. 2010).

The post-eruption of Mount Kelud in 2014 has a significant impact on the building material sector; a large number of river flows upstream on the peak area of the Kelud volcano by carrying large amounts of volcanic material resulting from eruptions has the potential to increase the population's economy (Wardhani et al. 2017). Agricultural conditions were developed by the people of Kediri and Blitar Regencies through planting pineapple plants. Ngancar Sub-district had made the pineapple Festival in October 2016 to introduce this plant to be an "icon" of the Ngancar sub-district. The local residents also dubbed the pineapple in this area a "Pineapple Queen." This is due to the fact that the planting area of this plant in the Ngancar district is more than 2,500 hectares and the production capacity is up to 50 tons/hectare. The tourism sector, which was originally only the peak of Mount Kelud, became several tourist destinations offered at the time of the post-eruption period in 2014. The development of tourist destinations in the lower slope area of Kelud Mountain is rapid as proven with the last three years of opening new tourist destinations located on the lower slope of Mount Kelud in the southwest, including Anggrek Village, Nanas Village in Ngancar District, Teletubies Hill, Karanganyar Coffee Plantation, Melon Village in Nglegok District, Durian Village in Plosoklaten District and Avocado in Puncu and Kepung Dirstrict.

The three significant impacts of the 2014 eruption are activities that are widely undertaken and developed even as a program from the local government as a development capital that emphasizes the wealth of local resources. Thus far, the study of the positive impact of the Mount Kelud eruption has been widely discussed, but the impact and response of local religious and cultural communities to the significance has not been studied. To fill such a void, the present study was designed to reveal fourfold issues: to what extent are local religious and cultural community responses around the slope of Mount Kelud against the significant impact and public understanding of environmental ethics in favor of preserving such significant natural potential through local wisdom and religiosity values, the significant impacts that provide significant benefits for the community and the negative impacts arising from the potential utilization of the landscape after eruption.

2 METHODS

In this study, the writer reveals that religion and local wisdom have a major role in the process of understanding the community around the slope of Mount Kelud against environmental ethics. The matters relating to religion and local wisdom are broad and can be grouped into two aspects: a visible aspect and an invisible aspect. Matters relating to religion will be referred to as theological constructions, given that their existence comes from religious teachings. In contrast, the things related to local wisdom can be called cultural constructions because they come from the values of the ancestral heritage of Javanese people on the slopes of Mount Kelud. Each of these constructions is concrete, meaning they can both be seen and observed directly, and there is also an abstract form, meaning it cannot be observed directly. It is stored in the community knowledge system of the slope of Mount Kelud and can only be known by the language phrases.

The major problem faced by human beings is the exploitation of natural and environmental resources. Among these cases, this study found that sand mining done by either by an individual or group has been pervasive. The impacts are catastrophic: loss of livelihood, social conflict, environmental pollution disruption of the landscape occur (Sturtevant 2009). This can also be found on the slopes of Mount Kelud. Sand mining done by one company has impacted the destruction of paddy fields and plantations. Looting the land is not only illegal but also eliminates damages plantation operations. Based on law No. 18 the year 2004 on plantations in part two of article 4, the plantation has several functions, such as: economic, which is an increase in prosperity and welfare of the community and the strengthening of the economic structure of the region and country; ecological, which is increase in soil and water conservation, carbon absorption, oxygen production and protected area buffers; socio-cultural, namely as strengthening relations and uniting nations. The occupation of the forest looting and deforestation has lead to the lost of PTPN XII function as the management of state plantations, such function was writen on the provision under LAW No. 18 year 2004. The case can be seen in *Afdeling Damarwulan Sepawon* Sub-district of Puncu, Kediri Regency The ecological function was vanished due to the logging of hundreds of trees in the location.

There are about 736 types of trees that are felled, consisting of Sengon, Balsa, Mahogany, Gmelina, Johar, Lamtoro, Jabon, Mindi and some other types. The area is a conservation area, where there is a water source flowing to the surrounding villages' residents (Firdaus et al. 2014). One of the impacts of conservation land excavation is the occurrence of landslides. Some parts of areas experience droughts because of water accumulating in certain regions, loss of populations' livelihoods, the destruction of residents dwellings due to massive mining activities and without tolerance of time and place, and sinkholes or the emergence of large holes in the soil as a result of excessive sand dredging. The phenomenon above does not mean to show that the surrounding community is not classified as not having a religion because Kediri is a base of Islamic boarding schools that examines the science of *Fiqh* (Indiyanto et al. 2012). Besides the practical books of *Fiqh*, many Islamic boarding schools in Kediri also examined the *Fiqh* nuances of Sufism, such as the *Bidayatul Hidayah* book by Imam Ghazali. Al-Ghazali noticed that Sufism and law had to be linked to spiritual life so that there was a close association. The function of public legal law is to regulate relationships between humans. The spiritual function is to discipline a person who has religion and purify his soul at the same time so that a combination of Fiqh and ethics lies in the pressures of heart and identity.

Arguments about the need for research to be done on environmental ethics in the area are departed from two things, first, because the study of *Fiqh* in an Islamic boarding school is still focused on the aspect of worship and the Mu'amalah (Masburiyah 2011). This statement was acknowledged by the An'im Falahuddin Mahrus, as quoted by Fakhruddin Mangunjaya, as a study of the environment has not been done in the Lirboyo Islamic boarding school, many of his discussions with students and Islamic scholars at Lirboyo are only related to societal issues, apparently, it seems that the environmental problems have not been studied (Mangunjaya 2007; 2014). Even Kyai An'im conveyed on another occasion afterward that environmental preservation is a determinant of natural balance. In the context of environmental preservation, this understanding has been heard for a long time. Even the lesson of natural science seems to unrelentingly teach that all components of the ecosystem, both tangible living beings, and other natural components, are united and must be balanced. However, in the layout of the application, the human must examine and question the effectiveness of the results of existing efforts. Obviously, after realizing such questions, human should introspect on various disaster portraits that have occorred in this hemisphere lately.

The attention to the aspect of environmental ethics that includes an ethical basis together for the people around the slope of Mount Kelud is important in preserving the environment and building public awareness to be committed together in environmental ethics on the slopes of Mount Kelud. Second, because the slope of Mount Kelud is demographically located by the area which is surrounded by Christianity, Buddhism, Hinduism and other religious traditions. According to religious adherent data published by the Office of the Ministry of Religion of Kediri Regency, the district and village areas located on the slope of Mount Kelud is a region that has more religions than Muslims when compared to other sub-districts, in the table the sub-districts that are located in the slope of Mount Kelud are: Wates, Ngancar, Plosoklaten, Gurah, Puncu, Kepung, Kandangan and Pare. The following table shows the number of religious adherents in the region (Lukens-Bull 2010).

From the data of environmental ethics among the religious people in the region, the sub-districts and the villages on the slope of Mount Kelud can build harmonious relationships between *interfaith* religions in forming a joint commitment in preserving the environment around the slope of Mount Kelud. Mount Kelud, for the surrounding community, has a symbolic meaning: first, concerning the aspect of social affairs. In the social context, Mount Kelud for the surrounding community economically gives the blessing of agricultural fertility as the livelihoods of thousands of farmers in the region, including the cultivation of fisheries with the abundant water from the source on the slopes of mountain Kelud (Anshoriy 2008). Therefore, the community of Mount Kelud has a strong bond of existence with Mount Kelud. In addition, Mount Kelud becomes an integral part of establishing the culture of the society. Social order is influenced by symbols that are used as the community's grip in daily life. Then, it will form traditions, values, norms, mythology in the survival of the everyday life of the people around Kelud.

The people around Kelud always carry out various ritual activities, usually lead to a friendly life cycle, such as the birth-related ceremony that began from the pregnancy to the ceremony before the birth of babies, circumcision and *pacangan*. With regard to death, the people of Kelud also held a traditional *Selametan* ceremony on days one, three, seven, fourteen, one hundred, and one year and 3 years after his death. Besides that, there are annual ceremonies, such as *Muludan*, *Rejeban*, *Nisu Sya'ban* (*Barakah* until late night), Ceremony of *Nyadran Bulan Ruwah*. The ceremonies' uniqueness lies in the rules of the food ingredients that are served, the place of ceremony, and its tools.

In the community belief tradition around Kelud, the older person always becomes a role model, and if they are dead, they are called ancestors. The term ancestor is always associated with the lineage that empties to the opening of the land (the forerunner of the village). The belief of the people around Kelud is the belief in many gods and is usually performed in *wayang* performances to convey educational and moral messages. In this case, two gods play an important role in society's life, namely the Goddess of fertility or *Dewi* Sri and *Dewa* Bathara Kala, who is believed to be able to resist the doom and misery of life.

In the community around Kelud, human interaction between social actors and cultural actors with volcanoes as a venue for socio-cultural activities has its background. Mount Kelud is an integral component of the surrounding community, both from the ecological system, the socio-cultural system, and territorial integrity. Despite the repeated disasters of Mount Kelud in Kediri and Blitar, the choice to settle in the volcanic area is not to be rethought. Generatively, the experience of interacting with the volcanic environment and making a community around Kelud can be friendly with the character of Mount Kelud, and they do not avoid it. Nevertheless, those living in the volcanic area remain wary that they live amid the disaster's potential insecurity.

In the Hindus' view, one form of intensive relationship with the mountain is to purify the mountain and pack it in a magical fashion by regularly worshipping the mountain ruler. The ruler of Kelud volcano is Hyang Acalapati or *Dewa* Acalapati, a mountain deity that existed only in Java and this deity also resides on Kelud. The kingdom of Kediri built Penataran Temple to worship the Lord of the Kelud so as not to anger him and to release his fire. In the era of the Majapahit kingdom, *Raja* Hayam Wuruk worshipped this deity in *Candi* Penataran. With the belief that Mount Kelud is a place of worship of the gods, society uses religious-magical mitigation to ward off fiery Kelud eruptions. This approach is more dominant than technological or engineering mitigation. This belief can be seen in Besowo village, Kepung District Kediri Regency. The people of this village still held ritual adoration of Kelud volcano. This adoration is known as *Argakerti* or mountain ceremonies.

This cult was created because the Kelud volcano had an important position. Kelud volcano is included in the *Pancamahabuta* or five important natural particles besides earth, oceans, lakes, and rivers. This cult was led by *Pandita Ida Pandita Wesnamatwaja*. The village of Besowo, whose majority of its citizens embraced Hinduism, always gave offerings in every *Argakerti* ceremony, such as the head of black buffalo and others. The *Argakerti* cult ceremonies are held as offerings to the *Sanghiang Giri Pati* or those who bring the mountain to life. In the Kelud volcano, there is also *Sanghiang Acalapati,* which is part of Kelud. The offerings to the two gods were intended so that the fire in Kelud would not flare up but also did not extinguish. In Penataran Temple held ceremonies worship for *Sanghiang Acalapati*, and in the village Besowo, District Kepung Kediri Regency, held ceremonies worship *Sanghiang Giri Pati*. The two ceremonies are done so that the Kelud volcano remains in a stable position.

People who are living in the area of *Gunung* Kelud have legendary folklore that is closely related to the existence of the mountain, such as; some people believe that the eruption occurred because of the anger of *Mahesa Sura* to the daughter of Kediri named the *Dewi* Kili Suci. The character is closely related to the legend of Panji as folklore that thrives in Kediri. It is said that Mahesa Sura, a powerful knight with a buffalo head fell in love and submitted a proposal to the *Dewi* Kili Suci. The proposal woukd be accepted by the princess if the suggestion was fulfilled, which was to make a large and deep well. *Mahesa Sura* was trapped and fell into the stone well. By the command of the princess, her troops pelted *Mahesa Sura* with a lot of rocks so that the well became a giant mound resembling a mountain. This story became the legend of Mount Kelud. Blitar people believe the story so that they think the eruption of Kelud will not significantly impact Blitar because the anger of *Mahesa Sura* leads to Kediri. *Wage Keramat*; wage is one of the market days in the Java Calendar system (*Pahing, Pon, Kliwon,* and *Legi*). People who believe in *Titen* (memory) science, believe that it is still valid in their daily lives and that Gunung Kelud will erupt on the day of the Wage market. *Mount Kelud* is a cemetery for the *Keris Empu Gandring*.

During the heyday of Majapahit Kingdom under the ruler of King Hayam Wuruk Gunung Kelud erupted. The eruption had a great impact and gained the attention of the king, since its impressiveness was considered able to muzzle the evil aura of the Kris Empu Gandring. During the Wisnuwardana period, the Keris was discarded in the crater Mount Kelud. The *Larung Saji* for rejecting disaster: People of Sugihwaras village, Ngancar district, Kediri Regency in each month of Suro always hold a ritual of the offering ceremony to reject the oath of *Mahesa Sura* (Lembu Sura). Some people also assume that this ceremony is a blessing form on the fertility of the Kelud Land (*subur gemah ripah loh jiwani*) and a tribute to the ruler of Mount Kelud. The Christian community in Gunung Kelud believes that *Tunggul Wulung* is legendary in Kediri and especially on Mount Kelud. It is believed to be the warlord of King Jayabaya who ruled the Kediri kingdom. In the narration of *Serat Babad Kadhiri,* written by *Mas Ngabehi* Poerbawidjaja and *Mas Ngabehi* Mangoenwidjaja, after completing his service for *Prabu* Jayabaya, Tunggul Wulung chose to be a demon that was commanded to stay and guard Mount Kelud against all dirty and wicked deeds.

In the current development of the Christian community in the slope of Mount Kelud, they build environmental ethics after the eruption with the establishment of the *Program Pengembangan Desa Bersaudara*, which means "The Village Brothers Development Program" or in commonly used the English words, "Sister Village." Which can simply can be interpreted as an effort to bring together a village that is in a disaster-prone area with villages that are in a safe area cooperate in disaster management. The movement can be found in the Sempu-Ngancar village, Segaran village, corner Wates village and Besowo village of the Kepung subdistrict. These villages contain the majority of Christians, namely Segaran village and the corner village of Wates District, and Damarwulan Village of Kepung Sub-district and Sidorejo village of Pare district. The term *Sister* is a popular phrase in the Dutch Colonial Hospital System that is often used by the church.

In the condition of *gugur gunung* or when the slope of Mount Kelud bestows the grace of potential natural resources, a sister village can be developed sustainability in developing cooperation and strengthening capacity and resources. For example, in the economics fields and foods, where one of the villages in an area that does not produce rice, it can cooperate with the production area and rice barn. Through BUMDes and farmer groups, the community can obtain good rice quality at affordable prices because it is directly from the farmer's production. Other advantages gained from this cooperation involve the farmer group and BUMDes, where much money circulates. Hence, the economic activity of the community can increase. The wider impact is that society can be more resilient food in the face of catastrophic potential (Yusron 2018).

4 CONCLUSION

This study has revealed the religiously and locally-based existence and meaning of mythology and rituals among Mount Kelud communities. Local community responses to the utilization of Gunung Kelud material can be found in various forms of collaboration between communities, such as various ritual activities that have always been carried out by the community around Kelud, usually based on a friendship life cycle. The community of Islamic boarding schools around Mount Kelud, represented explicitly by Lirboyo Islamic boarding school and other surrounding schools, has conducted an in-depth study of the classical *Fiqh* books implemented in the ethics environment. Hindu Society implements a form of intensive relationship with the mountain to cleanse the mountain and pack it magically, by regularly holding worship of the mountain rulers.

REFERENCES

Anshoriy, Nasruddin M. Ch. 2008. *Kearifan lingkungan dalam perspektif budaya Jawa*. Jakarta: Yayasan Obor Indonesia.

Firdaus, Maulana, Radityo Pramoda, & Maharani Yulisti. 2014. Dampak Letusan Gunung Kelud Terhadap Pelaku Usaha Perikanan Di Kabupaten Kediri, Provinsi Jawa Timur. *Jurnal Kebijakan Sosial Ekonomi Kelautan dan Perikanan* 4, no. 2: 157.

Haerani, Nia, Hendrasto, M., & H. Z. Abidin. 2010. Deformasi Gunung Kelud Pascapembentukan Kubah Lava November 2007. *Indonesian Journal on Geoscience* 5, no. 1: 13–30.

Indiyanto, Agus, & Arqom Kuswanjono, ed. 2012. *Kajian integratif ilmu, agama, dan budaya*. Cet. 1. Seri agama dan bencana. Bandung: Kerja sama Mizan Pustaka [dan] Program Studi Agama dan Lintas Budaya, Sekolah Pascasarjana, Universitas Gajah Mada, Yogyakarta.

Kelman, Ilan, & Tamsin A. Mather. 2008. Living with Volcanoes: The Sustainable Livelihoods Approach for Volcano-Related Opportunities." *Journal of Volcanology and Geothermal Research* 172, no. 3–4: 189–98.

Lukens-Bull, Ronald. 2010. Madrasa by Any Other Name: Pondok, Pesantren, and Islamic Schools in Indonesia and Larger Southeast Asian Region. *Journal of Indonesian Islam* 4, no. 1: 1-21–21.

Mangunjaya, Fachruddin M. 2014. *Ekopesantren: bagaimana merancang pesantren ramah lingkungan?* Cetakan pertama. Jakarta: Yayasan Pustaka Obor Indonesia.

Mangunjaya, Fachruddin M., ed. 2007. *Menanam sebelum kiamat: Islam, ekologi, dan gerakan lingkungan hidup*. Ed. 1. Jakarta: Conservation International Indonesia: Islamic College for Advanced Studies.

Masburiyah. 2011. Konsep Dan Sistimatika Pemikiran Fiqih Sufistik Al-Ghazali." *Nalar Fiqh* 4, no. 1.

Nyoman, Arsana. 2014. *Dedikasi TNI dalam bencana Sinabung dan Kelud*, Jakarta: Pusat Sejarah, Markas Besar Tentara Nasional Indonesia.

Sturtevant, William C. 2009. Studies in Ethnoscience1. *American Anthropologist* 66, no. 3: 99–131.

Tjandra, Kartono. 2015. *Mengenal gunungapi: bencana dan manfaat hasil letusannya*. Cetakan pertama. Yogyakarta: Gadjah Mada University Press.

Wardhani, Puspita Indra, Junun Sartohadi, dan Sunarto Sunarto. 2017. Dynamic Land Resources Management at the Mount Kelud, Indonesia. *Forum Geografi* 31, no. 1: 56.

Yusron. 2018. *Menguak pesona Gunung Kelud*, 1st edition. Yogyakarta: deepublish.

Author index

Milton Keynes UK
Ingram Content Group UK Ltd.
UKHW051925141024
449569UK00027B/1367